Springer Theses

Recognizing Outstanding Ph.D. Research

For further volumes:
http://www.springer.com/series/8790

Aims and Scope

The series "Springer Theses" brings together a selection of the very best Ph.D. theses from around the world and across the physical sciences. Nominated and endorsed by two recognized specialists, each published volume has been selected for its scientific excellence and the high impact of its contents for the pertinent field of research. For greater accessibility to non-specialists, the published versions include an extended introduction, as well as a foreword by the student's supervisor explaining the special relevance of the work for the field. As a whole, the series will provide a valuable resource both for newcomers to the research fields described, and for other scientists seeking detailed background information on special questions. Finally, it provides an accredited documentation of the valuable contributions made by today's younger generation of scientists.

Theses are accepted into the series by invited nomination only and must fulfill all of the following criteria

- They must be written in good English.
- The topic of should fall within the confines of Chemistry, Physics and related interdisciplinary fields such as Materials, Nanoscience, Chemical Engineering, Complex Systems and Biophysics.
- The work reported in the thesis must represent a significant scientific advance.
- If the thesis includes previously published material, permission to reproduce this must be gained from the respective copyright holder.
- They must have been examined and passed during the 12 months prior to nomination.
- Each thesis should include a foreword by the supervisor outlining the significance of its content.
- The theses should have a clearly defined structure including an introduction accessible to scientists not expert in that particular field.

Christian Spickermann

Entropies of Condensed Phases and Complex Systems

A First Principles Approach

Doctoral Thesis accepted by
Wilhelm-Ostwald-Institute of
Physical and Theoretical Chemistry,
University of Leipzig, Germany

Author
Dr. Christian Spickermann
Chair II of Inorganic Chemistry
Organom
Ruhr-University Bochum
Organometallics and Materials
Universitätsstraße 150
44801 Bochum
Germany
e-mail: christian.spickermann@rub.de

Supervisor
Prof. Dr. Barbara Kirchner
Wilhelm-Ostwald-Institute of Physical
and Theoretical Chemistry
University of Leipzig
Linnéstraße 2
04103 Leipzig
Germany
e-mail: bkirchner@uni-leipzig.de

ISSN 2190-5053 e-ISSN 2190-5061

ISBN 978-3-642-15735-6 e-ISBN 978-3-642-15736-3

DOI 10.1007/978-3-642-15736-3

Springer Heidelberg Dordrecht London New York

Cover design: eStudio Calamar, Berlin/Figueres

Printed on acid-free paper

Springer is part of Springer Science+Business Media (www.springer.com)

Parts of this thesis have been published in the following journal articles:

W. Reckien, C. Spickermann, M. Eggers, B. Kirchner, "Theoretical energetic and vibrational analysis of amide-templated pseudorotaxanes", *Chemical Physics*, **343**, 186–199 (2008)

C. Spickermann, T. Felder, C. A. Schalley, B. Kirchner, "How can rotaxanes be modified by varying functional groups at the axle? A combined theoretical and experimental analysis of thermochemistry and electronic effects", *Chemistry—A European Journal*, **14**, 1216–1227 (2008)

C. Spickermann, S. B. C. Lehmann, B. Kirchner, "Introducing phase transitions to quantum chemistry—From Trouton's rule to first principles vaporization entropies", *Journal of Chemical Physics*, **128**, 244506 (2008); *Virtual Journal of Biological Physics Research*, **16**, 1 (2008)

S. B. C. Lehmann, C. Spickermann, B. Kirchner, "Quantum cluster equilibrium theory applied in hydrogen bond number studies of water. Part I: Assessment of the quantum cluster equilibrium model for liquid water", *Journal of Chemical Theory and Computation*, **5**, 1640–1649 (2009)

S. B. C. Lehmann, C. Spickermann, B. Kirchner, "Quantum cluster equilibrium theory applied in hydrogen bond number studies of water. Part II: Icebergs in a two-dimensional water continuum?", *Journal of Chemical Theory and Computation*, **5**, 1650–1656 (2009)

B. Kirchner, C. Spickermann, W. Reckien, C. A. Schalley, "Uncovering individual hydrogen bonds in rotaxanes by frequency shifts", *Journal of the American Chemical Society*, **132**, 484–494 (2010)

Supervisor's Foreword

Theoretical investigation of complicated systems should include the treatment of the condensed phase, because it is mainly the liquid state in which most chemical processes occur. Traditional methods usually applied to study chemical phenomena encompass molecular dynamics (MD) simulations based on force fields as well as static quantum chemical (QC) methods. While both these methods have matured to become powerful tools over several decades, they almost complementarily share advantages and disadvantages. MD is good in allowing for large samples of molecules as well as long simulations times to be treated, while QC provides an excellent insight into the electronic structure of molecules as well as into their intermolecular forces. The disadvantages are clear; whereas MD neglects cooperativity and nuclear quantum effects, QC treats isolated molecules at zero temperature. Alternatives to these methods are given by ab initio molecular dynamics (AIMD) simulations and the quantum cluster equilibrium (QCE) method which was developed by Frank Weinhold [Weinhold, F. J. Chem. Phys. 1998, 109, 367]. As a combination of quantum chemistry with simulations, ab initio molecular dynamics simulations have the advantage to describe spontaneous events and important intermolecular forces as cooperativity. However, simple electronic structure methods have to be applied in AIMD and due to that AIMD is only able to perform in a range of quality prescribed by the electronic structure method which is combined with it. That only middle-sized ensembles can be calculated together with the short simulation times is the downside of a method which on the other hand allows one to obtain the electronic structure on the fly. The quantum cluster equilibrium method treats many clusters in a mean field approach and via the rigid rotor harmonic oscillator (RRHO) model allows one to derive partition functions for different phase points from high level ab initio QC (Kirchner, B. J. Chem. Phys. 2005, 123, 204116). With the partition function, the simple QCE method opens the access to thermodynamic quantities for a wide range of temperatures and pressures. Whereas cooperativity and nuclear quantum effects can be accounted for, there is no dynamics in the method and all the approximations made within the RRHO model might worsen the results. Thus, it becomes apparent that for the study of the microscopic details of complex chemical processes in large systems and in condensed phase, i.e. under conditions that closely resemble laboratory conditions, more accurate methods and at the same time faster methods are needed. An all-encompassing picture can be obtained only by scale-transferring concepts.

Theoretical investigation of complicated systems should also involve the calculation of quantities which can be compared to experimental findings. In seven chapters Christian Spickermann's thesis deals with the calculation of thermodynamic quantities in general and with entropy calculations especially. The manipulation of equations in order to arrive at the right form to be implemented in our computer programs (http://www.uni-leipzig.de/~quant/qce/pm_en.html) was one of the major goals of Christian Spickermanns thesis, next to the derivation of correction terms for the widely used equations to calculate thermodynamic quantities. Thereby important rules for the calculation of condensed phase thermodynamics from quantum chemistry have been established. However, the main reason for recommending his thesis for publication in this series is that Spickermann builds up the knowledge of theoretical possibilities from simple models to more complicated tools such as the QCE method in the most systematic and thus didactically effective way, from the treatment of isolated molecules to the one of the condensed phase. After a brief introduction, Spickermann explains each method from basics and discusses the approximations together with their advantages and disadvantages in order to allow the readers to derive their own conclusions. He carefully guides us through the methods and thereby makes it possible to gain from his experience which methods should be applied in which situation. This is followed by some case studies employing the simple RRHO model and the helpful discussion of the problems involved by using the example of the pseudorotaxane system as synthesized by Vögtle and Hunter [Hunter, C.A. J. Am. Chem. Soc. 1992, 114, 5303; Schalley et al. Acc. Chem. Res. 2001, 34, 465]. In the spirit of microsolvation, Christian Spickermann shows that simple but elegant concepts like describing a different kind of reaction (exchange with one solvent molecule instead of the basic association reaction) already leads to large improvements of the accuracy of calculated thermodynamic quantities. In two further chapters he offers insight into the outcome of QCE calculations based on different levels of quantum chemical accuracy and based on different levels of intercluster interaction. The associated liquids subject to his investigations are water and hydrogen fluoride, both complicated substances forming hydrogen bonds. He discusses cooperativity and its neglect as inherent in the pairwise additivity approximation of traditional molecular dynamics simulations. He shows that the thermodynamics of phase transitions is controlled by the stability of the clusters which rely on the accuracy of the chosen quantum chemical calculation. Error compensation as given by density functional theory in the standard formulation which is widely used in AIMD plays the same role in the liquid phase as in the gas phase. The influence of basis sets is analyzed. However, next to these important technical details, the choice of clusters plays a very important role. Distinct structural motives within the clusters are necessary in order to provide the best theoretically derived liquid phase. Given that these ingredients are chosen correctly, the QCE method is able to provide densities within chemical accuracy. It is also a large improvement over the simple RRHO model. Entropies can be calculated within 10 J/(mol K) and the transition entropies deviate less than 10% from the experimental values. Therefore, the present book does not only document a succesfull thesis, but should, in addition to this, serve as a demonstration of how systematic improvement of the calculated thermodynamics is possible when using highly accurate quantum chemical data and applying models and methods ranging from the RRHO model to the QCE method without and with intercluster interaction. Critical discussion of the results here accompanies the data evaluation so that calculation of condensed phase thermodynamics is conducted based on the first-principles idea of quantum chemistry. The benefits from this ansatz are clear: If we know which approximation and improvement leads to which particular result, we learn more about the nature of these chemical systems.

Leipzig, September 2010 Barbara Kirchner

Acknowledgments

Even if a doctoral thesis is written by a single person, it is in the end a joined effort of all the people who contributed to this work either by supervision or by support in a scientific as well as in a non-scientific sense. Here I want to express my gratitude to all these people and apologize for anyone who I might have forgotten in this context. The first entry on my list of acknowledgments is reserved for Barbara Kirchner, who supervised this thesis during the past three years. I want to thank her for financial support, for the wonderful atmosphere in the group and the scientific creativity, as well as for the many stimulating cooperations with national and international scientists, which resulted in interesting contacts around the globe. Special thanks also go to the students and colleagues who cooperated with me in many different projects during the last years and in that way contributed to the success of this thesis. These are (in alphabetic order): Michael v. Domaros, Sebastian B. C. Lehmann, Eva Perlt, Werner Reckien, Martin Roatsch, and Frank Uhlig. I also want to thank Jens Thar and Stefan Zahn for the cooperation regarding the joint publications and Miri Kohagen, Martin Brehm, and Matthias Schöppke for the really good time in the "party office". Large parts of this thesis could not have been realized without the aid and support of our scientific collaborators (in alphabetic order): Debra Bernhardt (Brisbane, Australia), Joachim Friedrich (Cologne, Germany), Markus Reiher (Zurich, Switzerland), and Christoph A. Schalley (Berlin, Germany), to all of which I want to express special acknowledgments. Furthermore, I acknowledge the financial support by the German Research Foundation (DFG) and the German Academic Exchange Service (DAAD). Last but not least I want to express my deep gratitude to my wife and my family, who accompanied me through the past three years with appreciation and supported me along the way. Thank you for the patience and the confidence which you have set in me all the time.

Bochum, September 2010 Christian Spickermann

Contents

List of Abbreviations and Symbols

aug–cc–pVTZ	Dunning's augmented, correlation-consistent valence triple-ζ basis set with polarization functions
aug–cc–pVQZ	Dunning's augmented, correlation-consistent valence quadruple-ζ basis set with polarization functions
BP	Density functional
B3LYP	Becke's three parameter hybrid functional
cbs	Complete basis set limit
CCSD	Coupled cluster including single and double excitations
CCSD(T)	CCSD with perturbative treatment of triple excitations
cp	Counterpoise correction
dft	(Kohn–Sham) density functional theory
fep	Free energy perturbation
fpmd	First principles molecular dynamics
fvci	Full vibrational configuration interaction
md	Molecular dynamics
MP2	Second order Møller-Plesset perturbation theory
fpmd	First principles molecular dynamics
NPA	Natural population analysis
PBE	Density functional
qce	Quantum cluster equilibrium
QZVP	Ahlrichs quadruple-ζ basis set
rrho	Rigid rotor harmonic oscillator
scf	Self-consistent field
SEN	Shared electron number
TZVP	Ahlrichs triple-ζ basis set
TZVPP	Ahlrichs triple-ζ basis set with extra polarization functions
E	Energy of N-particle system
ϵ	Energy of single particle
h	Planck constant
$\hbar$	Reduced Planck constant

$\mathcal{H}$	Hamilton function
$\hat{H}$	N-particle Hamilton operator
$\hat{h}$	Single particle (molecular) Hamilton operator
i_j	Number of monomers in cluster j
k_B	Boltzmann's constant
$\mathcal{K}$	Kinetic energy function
$\hat{K}$	Kinetic energy operator
λ	Wavelength
m	Mass of particle or nuclei
N_A	Avogadro constant
n_j	Number of monomers in cluster j
ρ_{n}	Number density
∇	Nabla operator
p	Pressure
P	Probability
$\mathbf{p}$	Momentum vector for N-particle system
Q	N-particle partition function
q	Single particle (molecular) partition function
$\mathbf{q}$	Coordinate vector for N-particle system
R	Gas constant
S	Entropy
T	Absolute temperature
t	Time
U	Interaction potential
V	Volume
W	Work

Chapter 1
Introduction

So far as physics is concerned, time's arrow is a property of entropy alone.

Sir Arthur Stanley Eddington [1]

Historically, classical thermodynamics can be understood as a phenomenological branch of the natural sciences dealing with the influence of heat and work on macroscopic systems. The dawn of quantum mechanics as well as the increase of computational resources led to the evolution and quantification of diverse microscopic perspectives on matter, which treat chemical systems as an ensemble of elementary particles and interactions among them. This thesis focusses on the "atomistic" scale of matter typically employed in chemistry, in which the fundamental units are the atomic nuclei as well as the electrons surrounding them. The availability of efficient software packages for the calculation of molecular properties on the basis of the various first principles methods of quantum chemistry during the last decades as well as the advancement in experimental techniques (for instance in laser spectroscopy) permits the chemical sciences to obtain detailed information in this microscopic time and length scale nowadays [2–5]. Accurate approximations to the molecular Schrödinger equation enable the calculation of molecular structures and the interactions between single molecules to high precision, and the advances in hardware technology as well as novel methodological developments permit the simulation of the condensed phase on the basis of first principles methods e.g. via the Car–Parrinello approach [6, 7]. However, this wealth of approaches to the atomistic scale in chemistry has not been available when the foundations of thermodynamics have been developed, and the elements of classical thermodynamics therefore do not correspond to this microscopic scale at first. The ideas and approaches finally closing the scale gap between the macroscopic quantities of classical thermodynamics and the microscopic scale of quantum mechanics are usually referred to as the theory of statistical mechanics nowadays, which dates back to the work of Boltzmann [8, 9]. The methods

C. Spickermann, *Entropies of Condensed Phases and Complex Systems*,
Springer Theses, DOI: 10.1007/978-3-642-15736-3_1,

of statistical mechanics provide a connection between the atomistic properties of individual molecules and the macroscopic thermodynamic quantities in terms of probability theory and the distribution of the particles constituting the macro-system over the available states. However, the application of these approaches for the calculation of thermodynamic properties of real chemical systems shows that the quantities necessary for the solution of the corresponding equations cannot be obtained in a straightforward way neither from computations nor experimental measurements, even though a direct link between quantum mechanics and thermodynamic quantities in principle does exist, see e.g. Chap. 2 of this thesis. In fact, the only "chemical" system which can be treated in an exact way by these approaches is the monoatomic ideal gas. Thus, the situation is not much different from quantum mechanics, where exact solutions can only be obtained for simple systems. Nevertheless, the developments of quantum chemistry in the recent decades demonstrate that the availability of exact solutions is no prerequisite for a successful application in the natural sciences if reasonable approximations can be found. The examination and evaluation of such approximations for the calculation of thermodynamic properties on the macroscale from the molecular properties on the microscale is the essential objective of this thesis. Thus, the investigated methods will intrinsically be multiscale approaches aiming at the calculation of thermodynamic state functions for the macroscopic N-particle system from the first principles of quantum mechanics. This course of action is fundamentally different to the concepts of empirical approaches for the computation of thermodynamic quantities like e.g. group contribution methods [10, 11]. These methods also employ the atomistic structure as the basic unit for the calculation of thermodynamic data, but in contrast to the first principles approaches examined in this thesis the only information relevant for the actual calculation is the molecular composition of the system under investigation. The final compound is additively constructed from smaller units (either single atoms or functional groups) which yield a constant contribution to the quantity to be calculated together with possible additional terms governing the interactions between different groups. The contributions of the different groups are obtained from experimental measurements of the corresponding pure compounds or appropriate mixtures with other substances. Thus, the real structural and energetic properties of the investigated system (like e.g. bond distances or dissociation energies) are not considered in these approaches, and the determination of thermodynamic quantities is made possible by applying sophisticated fitting procedures involving large data sets for the different groups. In this way properties of many different substances can easily be calculated to high precision, which is an important aspect e.g. for industrial applications. In contrast, the methods investigated in the present thesis might be less accurate and additionally depend on time-consuming quantum chemical calculations, but they arise from the first principles of quantum mechanics and statistical mechanics and thereby constitute true multiscale approaches by bridging the scale gap in consistency with these theories and not in terms of empirical correlations.

The investigations presented in this thesis set a particular focus on methods and approximations for the calculation of the entropy. In many chemical applications, the change in free energy constitutes the essential thermodynamic information, which in general permits the prediction of the thermodynamic feasibility of the process under investigation. As pointed out in Chap. 3, the straightforward application of the ideal gas model to more realistic situations (increasing degree of intermolecular interactions and higher densities) indicates that changes in the enthalpy in general are well-captured by this approach, whereas problems appear in the case of the entropy. Thus, methods for the determination of entropies in chemically realistic systems on the basis of quantum chemical first principles approaches constitute an important part of the theoretical branch in the molecular sciences (which can also be seen in the literature review sections of Chap. 3), and this thesis aims at contributing to the development and improvement of such methods.

References

1. Sir Eddington AS (1928) The nature of the physical world. The Macmillan Company, New York
2. Neese F, Wennmohs F (to be published 2010)
3. Ahlrichs R, Bär M, Häser M, Horn H, Kölmel C (1989) Chem Phys Lett 162:165–169
4. Neugebauer J, Herrmann C, Reiher M (2007) SNF 4.0. ETH Zürich, Zürich
5. Aßmann J, Kling M, Abel B (2003) Angew Chem Int Ed 42:2226–2246
6. Helgaker T, Jørgensen P, Olsen J (2004) Molecular electronic-structure theory. Wiley-VCH, Chichester
7. Car R, Parrinello M (1985) Phys Rev Lett 55:2471–2474
8. Boltzmann L (1995) Lectures on gas theory. Dover, New York
9. Boltzmann L (2000) Entropie und Wahrscheinlichkeit (1872–1905). Verlag Harri Deutsch, Frankfurt a. M.
10. Fredenslund A, Jones RL, Prausnitz JM (1975) AIChE J 21:1086–1099
11. Fredenslund A, Gmehling J, Rasmussen P (1977) Vapor–liquid equilibria using UNIFAC-A group contribution method. Elsevier, Amsterdam

Chapter 2
From Atomistic Calculations to Thermodynamic Quantities

This chapter gives a short summary of well-established as well as more recent pathways for extracting information about thermodynamic equilibrium quantities out of microscopic calculations, which are based on different approximations of quantum mechanics and as such only provide information about mechanical properties at first. The standard rigid rotor harmonic oscillator (rrho) model for the prediction of thermodynamic gas phase properties out of molecular quantities as well as its basis, the factorization of the N-particle partition function, will be revisited in detail along with a brief inspection of the different approximations this approach relies on. The quantum cluster equilibrium (qce) model will be exposed subsequently in terms of a van der Waals-like extension of the rrho approach for the thermodynamic treatment of condensed phases. In addition, several methods for computing thermodynamic equilibrium properties from molecular dynamics (md) simulations will be covered as well.

2.1 The Rigid Rotor Harmonic Oscillator Model

2.1.1 Essentials

2.1.1.1 The Classical N-Particle Partition Function

When dealing with the calculation of thermodynamic quantities from the fundamental molecular interactions, the ideas and methodologies of statistical thermodynamics are essential. The central quantity in statistical equilibrium thermodynamics is the partition function Q, which for a classical system of N identical and indistinguishable particles of mass m at temperature T confined to a volume V is given as [1]

C. Spickermann, *Entropies of Condensed Phases and Complex Systems*,
Springer Theses, DOI: 10.1007/978-3-642-15736-3_2,

$$Q_{\text{class}} = \frac{1}{N!h^{3N}} \int \exp(-\beta \mathcal{H}(\mathbf{p}, \mathbf{q})) d\mathbf{p} d\mathbf{q}. \tag{2.1}$$

In Eq. 2.1, β denotes the inverse temperature ($\beta = (k_B T)^{-1}$) and **q, p** are vectors containing generalized coordinates and the corresponding conjugate momenta, respectively, of all particles in the system (i.e., in the case of Cartesian coordinates $d\mathbf{p} = dp_{1x}dp_{1y}dp_{1z}...dp_{Nx}dp_{Ny}dp_{Nz}$ and $d\mathbf{q} = dx_1dy_1dz_1... dx_Ndy_Ndz_N$). The factor $N!^{-1}$ accounts for the redundancy of microstates introduced by the indistinguishability of the particles (for a detailed explanation see the part about the quantum mechanical limit in the following). In order to make the partition function dimensionless h has to be a normalization of the dimension momentum times length, which is taken to be the Planck constant, thereby ensuring consistency with the quantum mechanical limit [2, 3].

The Hamilton function represents the total energy of the system under examination and consists of a kinetic contribution $\mathcal{K}(\mathbf{p}) = \frac{1}{2m}\sum_{i=1}^{N}(p_{i,x}^2 + p_{i,y}^2 + p_{i,z}^2)$ and a configurational part $U(\mathbf{q})$[1]

$$\mathcal{H}(\mathbf{p}, \mathbf{q}) = \mathcal{K}(\mathbf{p}) + U(\mathbf{q}). \tag{2.2}$$

The presumably most simple many-particle system is realized by a system consisting of non-interacting particles, i.e., an ideal gas. In this case $U(\mathbf{q}) = 0$, and due to the independence of the momenta of different particles the integral in Eq. 2.1 can be evaluated directly, yielding [1]

$$\begin{aligned} Q_{\text{class,ideal}} &= \frac{1}{N!h^{3N}} \int \exp(-\beta K(\mathbf{p})) d\mathbf{p} d\mathbf{q} \\ &= \frac{1}{N!h^{3N}} \int \exp\left(-\frac{\beta}{2m}\sum_{i=1}^{N} p_{i,x}^2 + p_{i,y}^2 + p_{i,z}^2\right) dp_{1x}...dp_{Nz}dx_1...dz_N \\ &= \left(\frac{2\pi m k_B T}{h^2}\right)^{3N/2} \frac{V^N}{N!}. \end{aligned} \tag{2.3}$$

In Eq. 2.3 each particle contributes three Gaussian-type integrals via its three momentum coordinates, and each of the N integrations over the spatial coordinates yields the volume of the container V.

The form of Eq. 2.3 suggests the possibility to decompose the N-particle partition function (Eq. 2.1) into contributions q from the single particles which constitute the N-particle system according to

[1] Please note that the system consists of identical particles, which is the reason for the uniform mass m occurring in the kinetic contribution.

$$Q_{\text{class,ideal}} = \frac{q^N}{N!}, \quad \text{where } q = \left(\frac{2\pi m k_B T}{h^2}\right)^{3/2} V. \tag{2.4}$$

Strictly speaking, Eqs. 2.3 and 2.4 are only valid in the absence of any interparticle interactions, but even in the case $U(\mathbf{q}) \neq 0$, the spatial and momentum coordinates are independent and a separation of the form

$$Q_{\text{class}} = \frac{1}{N!}\left(\frac{2\pi m k_B T}{h^2}\right)^{3N/2} \times Z,$$
$$\text{where } Z = \int \exp(-\beta U(x_1, \ldots, z_N)) dx_1 \ldots dz_N, \tag{2.5}$$

is possible. In this equation, Z is called the *configurational integral* of the system, which depends only on the interparticle interactions. However, it is because of Z that the partition function of an N-particle system in most cases cannot be evaluated analytically, since a decoupling of the coordinates in $U(\mathbf{q})$ is normally impossible.

The separability of the N-particle partition function in single particle contributions in the absence of interparticle interactions is a quite general result and constitutes the basis for the rrho approach.

2.1.1.2 The Quantum Mechanical Limit

The quantum mechanical analogon to the classical Hamilton function $\mathcal{H}$ is given by the Hamilton operator, which for an N-particle system of identical particles takes the form [1]

$$\hat{H} = \hat{K} + U = -\frac{\hbar^2}{2m}\sum_{i=1}^{N} \nabla_i^2 + U(x_1, \ldots, z_N) \tag{2.6}$$

If $\{|\Psi_j\rangle\}$ denotes a set of eigenfunctions of $\hat{H}$, the Hamilton operator satisfies the Schrödinger equation

$$\hat{H}|\Psi_j\rangle = E_j|\Psi_j\rangle, \tag{2.7}$$

where E_j denotes the energy eigenvalue corresponding to the state the eigenfunction $|\Psi_j\rangle$ describes. Carrying out a one-to-one substitution according to Eq. 2.1 results in an expression of the form $\exp(-\beta\hat{H})$, which can be understood as a function of an operator. The effect of $\exp(-\beta\hat{H})$ on an eigenfunction of $\hat{H}$ can be illustrated by employing the corresponding MacLaurin expansion [1]

$$\exp(-\beta\hat{H})\,\Psi_j\rangle = \left(\sum_{n=0}^{\infty}\frac{(-\beta)^n}{n!}\hat{H}^n\right)|\Psi_j\rangle = \sum_{n=0}^{\infty}\frac{(-\beta)^n}{n!}\hat{H}^n|\Psi_j\rangle$$
$$= \sum_{n=0}^{\infty}\frac{(-\beta)^n}{n!}E_j^n\Psi_j = \exp(-\beta E_j)|\Psi_j\rangle, \tag{2.8}$$

which is again an eigenvalue equation. Solving for the jth eigenvalue, one obtains

$$\exp(-\beta E_j) = \langle\Psi_j|\exp(-\beta\hat{H})|\Psi_j\rangle, \tag{2.9}$$

since $\hat{H}$ is a hermitian operator and the $\{|\Psi_j\rangle\}$ can be normalized to an orthonormal set.

After transferring the integrand of the classical N-particle partition function, one has to take care of the integration itself, since the states of a quantum mechanical system are not continuously distributed over its state space. The classical partition function in Eq. 2.1 can be understood as a volume integral over the exponentially weighted energies the N-particle system can adopt. However, according to Eq. 2.7 a quantum mechanical N-particle system has a discrete set of energy values and there is no way to populate the state space between these allowed states, which is the reason why the integration in Eq. 2.1 can be replaced by a corresponding sum over the allowed states according to

$$Q_{\mathrm{qm}} = \sum_j \exp(-\beta E_j) = \sum_j\langle\Psi_j|\exp(-\beta\hat{H})|\Psi_j\rangle = \mathrm{tr}([\exp(-\beta\hat{H})]). \tag{2.10}$$

Since the trace of a linear operator's matrix representation is independent of the basis, the set $\{|\Psi_j\rangle\}$ is not particular in any form and Eq. 2.10 is a most general result as in case of Eq. 2.1, where in analogy the generalized coordinates **q, p** are not restricted to any special set [1].

According to Eq. 2.10, it is possible to calculate the partition function of a quantum mechanical N-particle system if either its energy eigenvalues or a basis of its state space is known. Despite this fact, Eq. 2.10 is of little practical use, since the Schrödinger equation Eq. 2.7 is exactly solvable only for one- and two-particle problems, and feasible approximations can be applied to systems of hundreds or thousands particles, which is far from the order of magnitude of macroscopic particle numbers ($\sim 10^{23}$) [4]. As in the classical case, a pragmatic approach to this size problem can be obtained by looking at an N-particle system without inter-particle interactions (i.e., an ideal quantum gas), since it is clear from Eq. 2.6 that the coupling between the particles is again solely due to the interaction potential $U(x_1, \ldots, z_N)$. Setting $U(x_1, \ldots, z_N) = 0$, the many-body Hamilton operator $\hat{H}$ can be decomposed into a sum of single particle contributions $\hat{h}_i$ [5]

$$\hat{H} = \sum_{i=1}^{N}\hat{h}_i, \quad \text{where } \hat{h}_i = -\frac{\hbar^2}{2m}\nabla_i^2. \tag{2.11}$$

Ignoring any possible symmetry constraints like the Pauli principle, the decoupling of the many-body Schrödinger equation Eq. 2.7 into single particle equations for this case can be realized via a wave function product ansatz of the form $|\Psi_j\rangle = |\psi_{j,1}\ldots\ \psi_{j,i}\ldots\ \psi_{j,N}\rangle$, where the set $\{\psi_{j,i}\}$ contains single particle eigenfunctions of the operators $\hat{h}_i$. The single particle Schrödinger equations thus take the form

$$\hat{h}|\psi_k\rangle = \epsilon_k|\psi_k\rangle, \tag{2.12}$$

where k labels the energy states of the single particle system. In this setup, every energy state E_j of the macrosystem can be expressed as a sum of the single particle energies $\{\epsilon_k\}$

$$E_j = \left(\epsilon_{k(1)}^{(1)} + \epsilon_{k(2)}^{(2)} + \cdots + \epsilon_{k(N)}^{(N)}\right) = \sum_{i=1}^{N} \epsilon_{k(i)}^{(i)}. \tag{2.13}$$

Each of the N identical particles will be in one of the single particle states labeled by k, i.e., the energy state can be understood as a function of the particle number, which is indicated by the label $k(i)$. Although all energy states E_j of the N-particle system can be written as sums of the single particle energies $\epsilon_{k(i)}^{(i)}$, this expansion is not unique in the case of identical particles, since a permutation of two or more indices will not change the resulting macrostate E_j. Consider for instance the expansions $\epsilon_1^{(1)} + \cdots + \epsilon_3^{(m)} + \epsilon_4^{(n)} + \cdots$, in which particle m is in energy state 3 and particle n is in energy state 4, and $\epsilon_1^{(1)} + \cdots + \epsilon_4^{(m)} + \epsilon_3^{(n)} + \cdots$, in which particle m is in energy state 4 and particle n is in energy state 3. Both sums finally lead to the same N-particle state E_j and therefore only one of them has to be included in an enumeration of the macrostates. For that reason it is more appropriate to assign occupation numbers $\{n_k\}$ to the single particle states $\{\epsilon_k\}$ in a system of identical particles and express the N-particle state according to [6]

$$E_j = \sum_k n_k \epsilon_k. \tag{2.14}$$

The values of n_k directly correspond to the number of particles occupying state ϵ_k, and the summation is now carried out over the single particle states labeled by k and no longer over the particles as in Eq. 2.13. Before evaluating the N-particle partition function in Eq. 2.10 with the aid of Eq. 2.14, one has to consider the correct weight for each of the macrostates E_j. As illustrated in the example above, it does not matter which of the individual particles occupies a given single particle state in a system of identical particles. This is the reason why a given macrostate E_j is completely characterized by the set of occupation numbers $\{n_k\}$. The correct weight g_j for the state E_j is thus given as

$$g_j(\{n_k\}) = \frac{N!}{n_1!n_2!\ldots} = \frac{N!}{\prod_k n_k!}, \tag{2.15}$$

which represents the number of possibilities to distribute the N identical particles with respect to the particular values of the occupation numbers $\{n_k\}$ constituting E_j according to Eq. 2.14 [7]. Applying these ideas, the N-particle partition function can be factorized in the following way (see also Sect. 7.2 in Chap. 7)

$$\begin{aligned} Q_{\mathrm{qm,ideal}} &= \sum_j g_j \exp(-\beta E_j) = \sum_{\{n_k\}} \frac{N!}{\prod_k n_k!} \exp\left(-\beta \sum_k n_k \epsilon_k\right) \\ &= \sum_{\{n_k\}} \frac{N!}{\prod_k n_k!} \prod_k \exp(-\beta n_k \epsilon_k) \\ &= N! \sum_{\{n_k\}} \prod_k \frac{[\exp(-\beta \epsilon_k)]^{n_k}}{n_k!}, \end{aligned} \tag{2.16}$$

where $\sum_{\{n_k\}}$ indicates the summation over all possible combinations of the occupation numbers n_k consistent with the condition $N = \sum_k n_k$. Equation 2.16 already represents a factorization into single particle contributions, but in general the summation over the sets of occupation numbers and the product over the energy states k cannot be exchanged as in case of the integration for a classical system, see Eq. 2.3.[2] In order to proceed further, a mathematical theorem known as the multinomial theorem has to be applied, which is given by [1, 7]

$$(x_1 + x_2 + \cdots + x_m)^c = \sum_{\alpha_1, \alpha_2, \ldots \alpha_m} \left(\frac{c!}{\alpha_1! \alpha_2! \ldots \alpha_m!}\right) x_1^{\alpha_1} x_2^{\alpha_2} \ldots x_m^{\alpha_m}. \tag{2.17}$$

As in Eq. 2.16, the summation over the set $\{\alpha_k\}$ is restricted according to $\sum_k \alpha_k = c$, and x_1, x_2, ... can be identified with the exponentials from Eq. 2.16. Comparing Eqs. 2.17 and 2.16, the final factorization is obtained as

$$\begin{aligned} Q_{\mathrm{qm,ideal}} &= \sum_{\{n_k\}} \prod_k \frac{N!}{n_k!} [\exp(-\beta \epsilon_k)]^{n_k} \\ &= [\exp(-\beta \epsilon_1) + \exp(-\beta \epsilon_2) + \cdots]^N \\ &= \left[\sum_k \exp(-\beta \epsilon_k)\right]^N \\ &= q^N, \quad \text{where } q = \sum_k \exp(-\beta \epsilon_k). \end{aligned} \tag{2.18}$$

Equation 2.18 demonstrates that one can calculate the partition function of a non-interacting N-particle system composed of identical particles from the single

[2] In fact, Eq. 2.16 does not represent any real simplification, because if the occupation numbers $\{n_k\}$ would be known for a macroscopic system, the N-particle state energies could be computed according to Eq. 2.14 and the N-particle partition function could be calculated in the conventional way, see Eq. 2.10.

particle partition functions q and the total number of particles N, thereby representing a real size reduction from the N-particle macrosystem to the single particle scale. However, by applying the weighting factor for the macrostates E_j from Eq. 2.15 one important restriction is tacitly assumed, namely the distinguishability of the N identical particles. If the N particles in turn are taken to be indistinguishable, there are no longer $N!$ different arrangements which could be distinguished from each other, but only one single arrangement. Therefore the transition from distinguishable to indistinguishable particles is formally achieved by a division by $N!$ as in the classical case, compare Eq. 2.3. The correct weight g_j of a given macrostate E_j for this case is thus given as [7]

$$g_j(\{n_k\}) = \frac{1}{n_1!n_2!\ldots} = \frac{1}{\prod_k n_k!}, \tag{2.19}$$

and the factorization of the N-particle partition function can be expressed as

$$\begin{aligned} Q_{\mathrm{qm,ideal}} &= \sum_{\{n_k\}} \prod_k \frac{1}{n_k!} [\exp(-\beta\epsilon_k)]^{n_k} \\ &= \frac{1}{N!} \left[\sum_k \exp(-\beta\epsilon_k) \right]^N \\ &= \frac{q^N}{N!}, \end{aligned} \tag{2.20}$$

where the single particle partition function q has the same meaning as in Eq. 2.18. An identical reduction was obtained for the classical setup (see Eq. 2.4), and factorization approaches of this kind constitute the first step towards the link between macroscopic thermodynamic state functions and atomistic calculations. However, up to this point Eqs. 2.18 and 2.20 are still of limited practical use, since in general there is an unlimited number of single particle states $\{\epsilon_k\}$ and the evaluation of the single particle partition functions q is not straightforward. In order to obtain a working relation for the computation of the N-particle partition function, approximations have to be introduced. Furthermore, in the treatment of chemically relevant systems one has to consider internal degrees of freedom for the particles as well as possible symmetry constraints on the N-particle wave function $|\Psi_j\rangle$, which have been omitted from the derivation up to now.

2.1.2 Molecular Systems and Approximations

2.1.2.1 Factorization of the Single Particle Partition Function

The conception presented in the last two parts emphasizes the formal difference between distinguishable and indistinguishable particles, but does not consider their internal structure. Even the simplest atomic systems will exhibit translational and electronic degrees of freedom, and in the case of molecular systems there will be rotational and vibrational contributions as well. The results of the last section show that in order to obtain the partition function Q for a non-interacting N-particle quantum system composed of indistinguishable and identical particles, it is sufficient to compute the single particle partition function q

$$q = \sum_k \exp(-\beta\epsilon_k), \tag{2.21}$$

where the phrase "particle" now stands for a single atom or even a molecule (i.e., a collection of elementary particles). The molecular energy states $\{\epsilon_k\}$ are solutions to the M-nuclei, n-electron Schrödinger equation, which, however, is not exactly solvable in general. Even if these solutions were available, there would be an unlimited number of them and the summation in Eq. 2.21 not necessarily has to converge to a well-defined limit. In order to tackle the problem, a series of approximations is usually applied, which, taken together, constitute the so-called rigid rotor harmonic oscillator approach [1]. The first of these approximations is the well-known Born-Oppenheimer approximation, which separates the molecular Schrödinger equation into two simpler equations, one for the electronic degrees of freedom (treating the nuclei as a constant external field) and one for the nuclear degrees of freedom (treating the electrons as a collective averaged potential) [5]. The Born-Oppenheimer approximation thus decouples the nuclear motion from the electronic motion, and the molecular Hamilton operator can be written as [8, 9]

$$\hat{h} = \hat{h}_{\mathrm{el}} + \hat{h}_{\mathrm{nuc}}, \tag{2.22}$$

where the electronic Hamilton operator $\hat{h}_{\mathrm{el}}$ includes the electron-nuclei interaction for a set of fixed positions of the nuclei and the nuclear Hamilton operator $\hat{h}_{\mathrm{nuc}}$ governs the motion of the nuclei.

The effect of an additive decomposition of the Hamilton operator as in Eq. 2.22 was already examined in the last section, compare Eqs. 2.11 and 2.13. For these equations it has been shown that the partition function of composite systems can be written as the product of the component partition functions if the Hamilton operator for the composite system is given as the sum over the Hamilton operators for each component, see Eq. 2.18. This general result can directly be applied to Eq. 2.22, which for the molecular partition function q yields a factorization of the form

$$q = q_{\mathrm{nuc}} q_{\mathrm{el}}. \tag{2.23}$$

Ignoring excited states of the nuclei (which are not populated in significant fractions at terrestrial temperatures), the remaining molecular degrees of freedom besides the electronic excitations arise from the translational motion of the whole molecule as well as rotations and vibrations. If there are no external fields present, the contributions entering the molecular Hamilton operator only depend on the relative distance between the nuclei [5]. In this case, the translational motion can always be separated from the other degrees of freedom by considering the translational motion of the center of mass and treating the internal degrees of freedom in a center of momentum frame, in which the molecular center of mass is at rest and only the relative motion of the nuclei is accounted for [1]. Since these motions are uncoupled, the nuclear Hamilton operator further reduces to

$$\hat{h}_{\text{nuc}} = \hat{h}_{\text{trans}} + \hat{h}_{\text{int}}, \tag{2.24}$$

where $\hat{h}_{\text{trans}}$ represents the Hamilton operator for the center of mass translation and $\hat{h}_{\text{int}}$ accounts for the remaining internal degrees of freedom, i.e., rotations and vibrations. The final decoupling of rotational ($\hat{h}_{\text{rot}}$) and vibrational ($\hat{h}_{\text{vib}}$) motion is rationalized on the basis of different timescales for these degrees of freedom as well as the observation that the vibrational amplitudes in many cases are small compared to the equilibrium distances between the nuclei, which justifies the treatment of rotations as being rigid [1, 10]. Following this reasoning, the nuclear Hamilton operator is given according to

$$\hat{h}_{\text{nuc}} = \hat{h}_{\text{trans}} + \hat{h}_{\text{int}} = \hat{h}_{\text{trans}} + \hat{h}_{\text{rot}} + \hat{h}_{\text{vib}}. \tag{2.25}$$

The final equations underlying the rigid rotor harmonic oscillator approximation for a quantum system composed of N identical and indistinguishable molecules can thus be summarized as [1]

$$\begin{aligned} \hat{h} &= \hat{h}_{\text{el}} + \hat{h}_{\text{trans}} + \hat{h}_{\text{rot}} + \hat{h}_{\text{vib}}; \quad \epsilon_{k(w,l,J,v)} = \epsilon_{w,\text{el}} + \epsilon_{l,\text{trans}} + \epsilon_{J,\text{rot}} + \epsilon_{v,\text{vib}}; \\ q &= q_{\text{el}} q_{\text{trans}} q_{\text{rot}} q_{\text{vib}} \\ &= \left[\sum_{w} \exp(-\beta\epsilon_{w,\text{el}})\right] \left[\sum_{l} \exp(-\beta\epsilon_{l,\text{trans}})\right] \\ &\quad \times \left[\sum_{J} \exp(-\beta\epsilon_{J,\text{rot}})\right] \left[\sum_{v} \exp(-\beta\epsilon_{v,\text{vib}})\right]; \\ Q &= \frac{\left[q_{\text{el}} q_{\text{trans}} q_{\text{rot}} q_{\text{vib}}\right]^N}{N!}, \end{aligned} \tag{2.26}$$

where the notation $\epsilon_{k(w,l,J,v)}$ again indicates the dependancy of the molecular state ϵ_k on the electronic state w, the translational state l, the rotational state J, and the vibrational state v. The set of approximations discussed above constitutes the main frame of the rigid rotor harmonic oscillator approach, but additional approximations will be necessary for the evaluation of the partition functions of the different degrees of freedom. These will be introduced in the following.

2.1.2.2 Translational Partition Function

According to Eq. 2.26 the translational contribution to the molecular partition function q takes the form

$$q_{\text{trans}} = \sum_{l} \exp\left(-\beta\epsilon_{l,\text{trans}}\right). \tag{2.27}$$

In order to evaluate this expression, an analytical form for the translational energy states $\{\epsilon_l\}$ has to be found. In the frame of the rrho approach, this is usually accomplished by treating the center of mass translation as the translation of a particle in a potential-free cubic box of length a with infinite potential walls at the borders of the box. This problem is a well-investigated quantum mechanical model system for which the analytic energy states are given as [11]

$$\epsilon(l_x, l_y, l_z) = \frac{h^2}{8ma^2}\left(l_x^2 + l_y^2 + l_z^2\right), \quad l_x, l_y, l_z \in \mathbb{N}^*, \tag{2.28}$$

where $\mathbb{N}^*$ denotes the natural numbers not including zero and m indicates the mass of the particle. The combination of Eqs. 2.27 and 2.28 results in

$$\begin{aligned} q_{\text{trans}}(V,T) &= \sum_{l_x,l_y,l_z}^{\infty} \exp\left(-\beta\epsilon(l_x,l_y,l_z)\right) = \sum_{l_x,l_y,l_z}^{\infty} \exp\left[-\frac{\beta h^2}{8ma^2}\left(l_x^2 + l_y^2 + l_z^2\right)\right] \\ &= \left[\sum_{l_x}^{\infty} \exp\left(-\frac{\beta h^2 l_x^2}{8ma^2}\right)\right]\left[\sum_{l_y}^{\infty} \exp\left(-\frac{\beta h^2 l_y^2}{8ma^2}\right)\right]\left[\sum_{l_z}^{\infty} \exp\left(-\frac{\beta h^2 l_z^2}{8ma^2}\right)\right] \\ &= \left[\sum_{l}^{\infty} \exp\left(-\frac{\beta h^2 l^2}{8ma^2}\right)\right]^3, \end{aligned} \tag{2.29}$$

where the last identity is valid due to the independence of the quantum numbers $\{l_k\}$ from each other. The sum occurring in the last line of Eq. 2.29 cannot generally be written in a closed form, which prevents a direct evaluation of q_{trans}. In order to obtain a working relation, an approximate treatment of the summation as an integration is normally done. This can be rationalized according to the following reasoning. The distance $\Delta\epsilon_l$ between adjacent translational states in most cases is very small compared to the thermal energy $k_{\text{B}}T$ at ambient temperatures. In these cases and at elevated temperatures, the discrete translational spectrum can thus be treated as a continuum, and therefore this approximation is referred to as the high temperature limit [11]. The same reasoning will again be employed in the discussion of the rotational partition function and in the consideration of symmetry constraints, see below. The application of that approximation results in a Gaussian-type integral, which can directly be evaluated according to

$$q_{\text{trans}} \approx \left(\int_0^\infty \exp\left[-\frac{\beta h^2 l^2}{8ma^2} \right] dl \right)^3 = \left(\frac{2\pi m k_B T}{h^2} \right)^{3/2} V, \tag{2.30}$$

where the volume V of the box is given as $V = a^3$. The result in Eq. 2.30 has already appeared before, namely as the single particle partition function in the classical treatment of an N-particle system of non-interacting particles, see Eq. 2.4. This is a plausible result, since the only single particle degrees of freedom in a system of non-interacting particles with no internal structure are expected to be translational degrees of freedom. It can also be seen that the ad hoc introduction of the Planck constant h in Eqs. 2.1 and 2.4 occurs in a natural way by treating translation as a quantum mechanical phenomenon in terms of the energy eigenvalues in Eq. 2.28.

A common practice in statistical thermodynamics is the introduction of the thermal de Broglie wavelength according to the following definition [1]

$$\Lambda = \left(\frac{h^2}{2\pi m k_B T} \right)^{1/2}. \tag{2.31}$$

Λ has units of length, and according to the classical equipartition theorem the average kinetic energy is given as $\langle \varepsilon_{\text{kin}} \rangle = p^2/(2m) = (3/2)k_B T$, i.e., the average momentum equals $(mk_B T)^{1/2}$. Inserting this average momentum in the conventional de Broglie relation $\lambda = h/p$ indicates the analogy to the definition of Eq. 2.31, and accordingly the translational partition function can be written as

$$q_{\text{trans}} \approx \frac{V}{\Lambda^3}. \tag{2.32}$$

The thermal de Broglie wavelength Λ can be used to estimate the importance of quantum effects and the applicability of classical approximations like the high temperature limit, i.e., a classical or semi-classical treatment is only reasonable if $\Lambda^3/V \ll 1$ [11].

2.1.2.3 Rotational Partition Function

In order to obtain expressions for the molecular rotational energy states, the quantum mechanical model system of a rigid rotor is employed. The energy eigenvalues for this problem are given by [1]

$$\epsilon_J = \frac{\hbar^2 J(J+1)}{2I}, \quad J \in \mathbb{N}, \tag{2.33}$$

where I indicates the molecular moment of inertia and J denotes the rotational quantum number. Equation 2.33 is only valid in this form if the three principal moments of inertia I_A, I_B, I_C of the molecule under study are identical, i.e., if the molecule belongs to the class of spherical tops [1]. The most general case of an asymmetric top ($I_A \neq I_B \neq I_C$) is a complicated problem and will not be treated in full detail here, but the quantum mechanical result for a spherical top can easily be generalized to the classical result for an asymmetric top in a plausible way. It is important to note that each of the energy states ε_J is degenerated by a factor $(2J + 1)^2$, which leads to the following form of the rotational partition function [1]

$$q_{\text{rot}}(T) = \sum_{J=0}^{\infty}(2J+1)^2 \exp\left(-\beta\frac{\hbar^2 J(J+1)}{2I}\right). \tag{2.34}$$

As in the case of the translational partition function, this sum cannot be expressed in any closed form, which is the reason why the summation again has to be approximated by an integration. The same reasoning leading to the high temperature limit in the case of the translational degrees of freedom can be applied for the rotational degrees of freedom as well, with possible exceptions to that given by molecular species showing small principal moments of inertia, for instance hydrogen. In this continuum limit, the rotational partition function can thus be written as [1]

$$q_{\text{rot}}(T) \approx \int_0^{\infty} (2J+1)^2 \exp\left(-\beta\frac{\hbar^2 J(J+1)}{2I}\right) dJ \approx \int_0^{\infty} 4J^2 \exp\left(-\beta\frac{\hbar^2 J^2}{2I}\right) dJ$$
$$\approx \pi^{1/2}\left(\frac{2Ik_{\text{B}}T}{\hbar^2}\right)^{3/2}, \tag{2.35}$$

where in addition the increased importance of large values of J at high temperatures is assumed, so that $J^2 \gg J \gg 1$.

The rotational symmetry has another important impact on q_{rot} besides the classification as asymmetric or spherical top, namely in the case of the true number of distinguishable rotational states. If there are any rotational symmetry elements in the molecular point group which transfer the molecule into itself (besides a C_1 axis), there will be an overcounting of identical rotational states. As in the case of $g(\{n_k\})$ (see Eq. 2.15), the correct weighting can be obtained by neglecting the redundant states, which is formally achieved in terms of the symmetry number σ. In the general case of a polyatomic molecule, this number equals the order of the rotational subgroup of the molecular point group, i.e., $\sigma = 2$ for water and $\sigma = 12$ for benzene. A more elaborate derivation of the symmetry number on the basis of the molecular wave function can be found in [1] (see also the section about symmetry constraints). The consideration of this symmetry effect results in the following general form of the rigid rotor (rr) partition function

$$q_{\text{rr}}(T) = \frac{\pi^{1/2}}{\sigma}\left(\frac{2Ik_{\text{B}}T}{\hbar^2}\right)^{3/2}. \tag{2.36}$$

The generalization of Eq. 2.36 to asymmetric top structures can be achieved by inserting the classical Hamilton function of a rigid rotor into the classical partition function (see Eq. 2.1) and carrying out the integration. The result is given as [1]

$$q_{\mathrm{rr}}(T) = \frac{\pi^{1/2}}{\sigma}\left(\frac{2I_a k_{\mathrm{B}} T}{\hbar^2}\right)^{1/2}\left(\frac{2I_b k_{\mathrm{B}} T}{\hbar^2}\right)^{1/2}\left(\frac{2I_c k_{\mathrm{B}} T}{\hbar^2}\right)^{1/2}$$
$$= \frac{\pi^{1/2}}{\sigma}\left(\frac{T^3}{\Theta_a\Theta_b\Theta_c}\right)^{1/2}, \tag{2.37}$$

where $\Theta_{(a,b,c)}$ denotes the rotational temperature associated with the principal moments of inertia I_a, I_b, I_c according to

$$\Theta_{(a,b,c)} = \frac{\hbar^2}{2I_{(a,b,c)}k_{\mathrm{B}}}. \tag{2.38}$$

It is apparent that this result reduces to the correct expression for a spherical top in the limit $I_a = I_b = I_c$.

Equation 2.37 represents a compact form of the rotational partition function and can be easily computed if the three principal moments of inertia are known. The central approximations employed in the derivation of Eq. 2.37 include the high temperature limit (Eq. 2.35) and the evaluation of the classical expressions in the case of asymmetric top structures.

2.1.2.4 Vibrational Partition Function

The quantum mechanical model system applied for the treatment of the molecular vibrational degrees of freedom is the model of the harmonic oscillator, for which the energy eigenvalues are given by [1]

$$\epsilon_v = \left(v + \frac{1}{2}\right)h\nu, \quad v \in \mathbb{N}. \tag{2.39}$$

Here ν denotes the oscillator's frequency, and for the case $v = 0$ the characteristic zero point energy $\epsilon_0 = \frac{h\nu}{2}$ of the oscillator is apparent from Eq. 2.39. According to Eqs. 2.26 and 2.39, the harmonic oscillator (ho) partition function q_{ho} can be expressed as

$$q_{\mathrm{ho}}(T) = \sum_{v=0}^{\infty}\exp\left(-\beta\left[v + \frac{1}{2}\right]h\nu\right)$$
$$= \exp\left(\frac{-\beta h\nu}{2}\right)\sum_{v=0}^{\infty}\exp(-\beta v h\nu)$$
$$= \frac{\exp(-\frac{1}{2}\beta h\nu)}{1 - \exp(-\beta h\nu)}, \tag{2.40}$$

where the last identity is obtained from the limit of the geometric series $\sum_{i=1}^{\infty} x^i = (1-x)^{-1}$ [11]. The harmonic oscillator partition function is the only one in the rrho approach which can be evaluated directly and without any approximation (compare the integral substitution in the case of the translational and rotational partition functions), but the treatment of molecular vibrations as harmonic oscillations is an approximation in itself. The generalization of Eq. 2.40 to the $(3M-6)$ vibrational degrees of freedom of a non-linear polyatomic molecule is straightforward and can be done in terms of the harmonic approximation. Following the reasoning from Sect. 2.1.2.1, the vibrational amplitudes are small compared to the overall molecular dimensions in most cases, and the vibrational motion of the M nuclei can thus be approximated as a movement in a harmonic potential around an equilibrium configuration [1]. The coupled $(3M-6)$-dimensional Schrödinger equation for this problem can be efficiently decoupled into $(3M-6)$ exactly solvable one dimensional harmonic oscillator Schrödinger equations via a normal mode analysis. The normal mode analysis is an orthogonal coordinate transformation which diagonalizes the Hessian, thereby eliminating the dependancy of a given vibration from all the other vibrations in the harmonic potential [4]. In the normal coordinate representation, the vibrational Hamilton operator $\hat{h}_{\text{vib}}$ is thus given as a sum of single mode harmonic oscillator Hamilton operators, and according to the reasoning leading to Eq. 2.18 the polyatomic vibrational partition function can be expressed as a product of the single mode contributions

$$q_{\text{vib}}(T) = \prod_{k=1}^{3M-6} q_{k,\text{ho}} = \prod_{k=1}^{3M-6} \frac{\exp(-\frac{1}{2}\beta h \nu_k)}{1-\exp(-\beta h \nu_k)}, \tag{2.41}$$

where ν_k denotes the frequency of the kth normal mode. According to Eq. 2.41, the calculation of the polyatomic harmonic oscillator partition function can be carried out if the normal mode frequencies $\{\nu_k\}$ are known. The normal mode analysis is a standard procedure of quantum chemistry and implemented in many quantum chemical codes [12, 13], so that the direct computation of q_{vib} is normally possible without any significant problems. However, one should keep in mind that Eq. 2.41 is derived from the harmonic approximation and that possible anharmonic corrections applied in the calculation of the frequencies also have to be considered in the computation of q_{vib}.

2.1.2.5 Electronic Partition Function

In complete analogy to the previous degrees of freedom, the electronic partition function is given as the sum over the molecular electronic ground state as well as all excited states according to

$$q_{\text{el}}(T) = \sum_{w} \exp(-\beta \epsilon_w). \tag{2.42}$$

There is no simple model system which could provide the molecular energy states ϵ_w as a function of the quantum number w. Different quantum chemical methods give analytic expressions for the energy eigenvalues of the electronic Schrödinger equation on the basis of the Born-Oppenheimer approximation, but in general these equations are quite complex and have to be solved iteratively [9]. Therefore, an exact calculation of the sum in Eq. 2.42 is not possible, but since the excited electronic states in most cases are separated from the ground state by energy gaps large compared to the thermal energy at ambient temperatures, it is generally sufficient to consider only the first few or even only the first term of the series. A commonly encountered problematic case for this approach are the halogen atoms, which show first excited states less than 0.1 eV above the ground state [1]. However, sample calculations show that even in these cases it is sufficient to consider only the first terms in Eq. 2.42 [11]. This approach in a way reverses the reasoning in the derivation of the translational and rotational partition functions and can be understood as a "reversed high temperature limit" as it gets more inaccurate as the temperature increases.

As in the case of the rotational degrees of freedom, a possible degeneracy of the electronic states has to be considered in the evaluation of the electronic partition function. If the wth energy state is degenerated by a factor g_w and the electronic ground state of the molecule is set to zero, the electronic partition function can be expressed as

$$q_{\mathrm{el}} = g_1 + g_2 \exp\left(-\beta \Delta\epsilon_{1,2}\right) + \cdots, \tag{2.43}$$

where $\Delta\epsilon_{1,2}$ denotes the energy difference between the first excited state and the ground state.

2.1.2.6 Symmetry Constraints and the High Temperature Limit

The quantum mechanical treatment of the non-interacting N-particle problem in Sect. 2.1.1 explicitly excluded possible symmetry constraints. However, all known particles are either classified as bosons (wave function of an identical N-particle system is symmetric under interchange of two particles) or fermions (wave function of an identical N-particle system is antisymmetric under interchange of two particles) [5]. The distribution of independent identical fermions over the single particle states is restricted insofar as no two fermions can occupy the same state, thereby introducing additional constraints to the composition of the N-particle energy states $\{E_j\}$ from the single particle states $\{\epsilon_k\}$, see Eqs. 2.13 and 2.14. Even in the case of bosons for which no occupation limit of a given single particle state exists, the correct enumeration of microstates is still not trivial in the case of indistinguishable particles. For both the classical as well as the quantum mechanical case, the transition from the distinguishable to the indistinguishable N-particle system in Sect. 2.1.1 is achieved by neglecting the number of permutations $N!$ which are no longer distinguishable in the case of indistinguishable particles.

However, the number of distinct permutations is only equal to $N!$ if each of the N particles occupies a different energy state ϵ_k, i.e., if the same condition as in the case of fermions is fulfilled. This can be seen most directly by considering the case where $N-1$ bosons occupy the same state (e.g. ϵ_m) and one boson occupies another state (e.g. ϵ_n). If the particles are distinguishable, there are N possbile choices for the particle occupying state ϵ_n, and the number of redundant permutations one would have to consider for the transition to indistinguishable particles is thus N and not $N!$. This clearly indicates that the division by $N!$ is only exact in the limiting case of individually occupied single particle states, the condition which also has to be satisfied in the case of an N-particle fermion system. Nevertheless, Eq. 2.20 is applicable in most cases due to the same reasoning already encountered in the derivation of the translational and rotational partition functions, namely the high temperature limit. At almost all conditions except the lowest temperatures and the highest densities, the number of available single particle states exceeds the number of particles N significantly, thereby ensuring that in most cases each particle will be in a different state and that the division by $N!$ is correct. These cases for which Eq. 2.20 is valid are said to obey Boltzmann statistics [1]. The underlying assumptions of Boltzmann statistics will become more and more probable with increasing temperature, e.g. for a particle of mass $m = 10^{-25}$ kg confined to a cubic box of 0.1 m length at $T = 300$ K the number of translational states alone is of the order of 10^{30} [1].

The reasoning presented so far indicates that Boltzmann statistics is applicable to most problems at ordinary conditions, and all further derivations and results presented in this thesis are based on this approximation. For the sake of completeness it is mentioned that the exact treatment of the symmetry constraint of bosonic and fermionic N-particle systems can be accomplished and that the resulting equations are referred to as Bose–Einstein statistics and Fermi–Dirac statistics, respectively [6].

2.2 The Quantum Cluster Equilibrium Approach

2.2.1 *Essentials*

2.2.1.1 The Ideal Cluster Gas

The rigid rotor harmonic oscillator (rrho) model presented in the last section constitutes a well-established approach for the calculation of thermodynamic quantities on the basis of atomistic calculations. However, since it is based on the factorization of the N-particle partition function Q into the molecular partition functions q which is only exact for a non-interacting system, the rrho model can be expected to be a reasonable approximation in the case of dilute gases only, for which the interparticle interaction is comparatively small. A possible first step for the extension of the standard rrho model to interacting systems can be realized by including interparticle interactions into the single particle unit for which the "single particle" partition function q is computed, i.e., by extending the molecular unit to a multi-molecular unit, which will be called a "cluster" from now on. Thus, in this way a cluster is defined as a molecular or atomic aggregate which is held together by some form of (attractive) interaction. This interaction could for instance be primarily coulombic in nature (e.g. in the case of sodium chloride clusters), but non-covalent interactions like hydrogen bonding (water clusters) or dispersion interactions (noble gas clusters) are thinkable as well. This approach transfers the interparticle interaction into the fundamental unit partly and in a local way. The cluster as the new fundamental unit will most generally be treated as the single molecule in the case of the conventional rrho approach, but in most cases there will be two different levels of interaction present within the cluster: the true *intramolecular* interactions which form the single molecule the cluster is composed of and which are most often covalent in nature, and the less strong *intermolecular* interactions, which are transferred into the cluster (thereby becoming *intracluster* interactions) and which ensure that the cluster as an aggregate of molecules will be stable. This classification is of course artificial: in the quantum chemical calculation of a cluster, only the type and position of the nuclei as well as the number of electrons enter, and no distinction between different molecules is made. It is also obvious that in the case of atomic clusters this classification is unnecessary, since only the interatomic interactions are present.

The combination of this new reference level with the ideas and approximations from the rrho approach is straightforward. Before, the fundamental units were molecules which have been constructed from atoms, thereby introducing rotational and vibrational degrees of freedom. Now, the fundamental units are clusters which are constructed from molecules, and since a cluster can be understood as a single "supermolecule", no additional degrees of freedom are necessary. Thus, in complete analogy to the molecular partition function from the last section a cluster partition function is now considered, which can be factorized according to the set of rrho approximations

$$q_j = q_{j,\mathrm{el}} q_{j,\mathrm{trans}} q_{j,\mathrm{rot}} q_{j,\mathrm{vib}}, \tag{2.44}$$

where j is a cluster label which will become important later on. The partition functions for the different degrees of freedom on the right hand side of Eq. 2.44 are evaluated according to the formulas derived in the last section, for which the corresponding cluster quantities are used (e.g. the mass of the cluster for $q_{j,\mathrm{trans}}$ and the frequencies of the cluster normal modes for $q_{j,\mathrm{vib}}$). This approach is straightforward and allows an easy computation of the cluster partition function q_j. However, one has to be careful in the case of the comparison between different sized clusters, for instance if the change of a thermodynamic variable for the reaction of two smaller clusters to a larger one is of interest. In such a case, the number of the different kinds of cluster degrees of freedom will change, and since the degrees of freedom are treated by different approximations as presented in the last section, there will be an artificial effect arising due to the different accuracy of the approximations employed. Consider for instance two water molecules which form a dimer cluster. Taken together, the two molecules have six translational and rotational degrees of freedom, but the dimer only has three of each of them. In contrast, the number of vibrations in the dimer is 12, whereas both molecules only possess 6 vibrational degrees of freedom. Thus, in the dimer three hindered translations and three hindered rotations will be treated as harmonic vibrations, and the question arises if the harmonic approximation is as accurate for hindered translations/rotations as the model of the particle in a box for free translations or the rigid rotor for free rotations. These questions will be examined in greater detail in Chap. 3.

If only clusters of a certain type j (e.g. of a certain size) are considered, the canonical partition function for a non-interacting n-particle system of these (indistinguishable) clusters in complete analogy to Eq. 2.20 is given as

$$Q_j(n_j, V, T) = \frac{q_j^{n_j}(V, T)}{n_j!}. \tag{2.45}$$

Even if all *intercluster* interactions (i.e., interactions *between* the clusters) are neglected in Eq. 2.45, it is certainly an improvement over the conventional rrho model in the modeling of a condensed system, since the *intermolecular* interactions of the true system are partly accounted for within the cluster via the quantum chemical calculation, and this fraction of interactions considered will become larger as the cluster size increases. An even more detailed approach would consider not only one certain cluster species j, but many different of them, which could resemble different structural patterns occurring in the real condensed phase one is trying to model. Among them one might also include the single *molecular* monomer, since this isolated structure could occur in the real system from time to time, especially in systems showing strong structural fluctuations. These different cluster structures $\{j\}$ will certainly be distinguishable from each other (e.g., by their size), and according to that the total N-particle partition function for an ideal mixture of these different cluster species is obtained as (see also Eq. 2.18) [14, 15]

$$Q(N, V, T) = \prod_j Q_j(n_j, V, T) = \prod_j \frac{q_j^{n_j}(V, T)}{n_j!}, \tag{2.46}$$

where n_j now denotes the *cluster population* of the corresponding cluster species j. To make sure that Eq. 2.46 is valid, one has to establish a relation between the total number of particles N and the cluster populations $\{n_j\}$. For instance, the total number of particles N could be equal to the number of *molecular* monomers in the system, which would be suitable for the calculation of thermodynamic quantities for, e.g., one mole of water molecules. In this case, the relation between N and the set $\{n_j\}$ is given by

$$N = \sum_j i_j n_j, \tag{2.47}$$

where i_j denotes the number of monomer units in the cluster species j. From the considerations presented so far, there is no simple way for the determination of the cluster populations $\{n_j\}$. As a starting option, the N molecular monomers could be distributed evenly among the different cluster species $\{j\}$, and from the partition function (Eq. 2.46) the free energy of the system could be obtained. Changing the distribution $\{n_j\}$ slightly and recalculating the free energy would then yield information about whether the change in population has a stabilizing or a destabilizing effect on the system. If there is information available on the relative abundance of a certain structural pattern over another (e.g., the relative stability of two modifications), these could be employed as well.

Even if the sketched approach will be superior to the conventional rrho model in most cases, the accurate modeling of condensed phases will have to include cluster–cluster interactions which will become more and more important at increasing densities. Furthermore, the volume of the particles will no longer be negligible at these conditions. An approximate treatment of these factors will be presented in the following.

2.2.1.2 The van der Waals Cluster Gas

The ideal cluster gas model derived in the previous part would in principle be able to treat a larger and larger fraction of interparticle interactions as the considered cluster sizes increase, but due to computational limitations there will be restrictions on the cluster sizes and therefore surface effects will always be present. A more convenient way for the treatment of additional interactions lies in a non-local mean field approach as realized in the model of van der Waals gases [11]. In this approach, the gas particles are no longer treated as point particles, but are assigned a non-zero size via a volume parameter b, and the (attractive) interparticle interaction is introduced in terms of a reduced pressure of the gas on the walls of the container which is governed by the mean field parameter a. Thus, an extension of the conventional rrho model on the basis of the van der Waals gas will have to consider the volume of the particles as a function of b as well as some form of interparticle mean field interaction as a function of a. The treatment of the reduced volume effect is achieved in a straightforward way by assigning a cluster volume V_j to each cluster species j and considering the occurrence of this cluster species in terms of the corresponding cluster populations n_j. The total excluded volume V_{ex} is thereby given as the weighted sum over all cluster species according to [14, 15]

$$V_{\mathrm{ex}} = b_{\mathrm{xv}} \sum_j n_j V_j, \tag{2.48}$$

where b_{xv} is a proportionality constant through which the cluster volume estimates V_j can be corrected in an average fashion. This is important, since the volume of a particle is no direct observable and there is no straightforward way of calculating the numbers $\{V_j\}$ [14]. In the present thesis as well as in many applications, the cluster volume V_j is estimated as the sum of the atomic sphere volumes of the different atoms contributing to the cluster j, which in turn are obtained from the corresponding van der Waals radii [16, 17]. A more elaborate treatment of the cluster volumes could be obtained in terms of the GEPOL algorithm [18]. The excluded volume term calculated in this way subsequently has to be employed for the correction of the overall volume available to the N-particle system. The approximations introduced in Sect. 2.1.2 demonstrate that the translational partition function $q_{j,\text{trans}}$ is the only contribution to $Q(N, V, T)$, which depends on the volume and that the dependancy is that of a simple linear relation (see Eq. 2.30). According to that, the translational partition function corrected for the excluded volume is given as [14, 15]

$$q_{j,\text{trans}} = \left(\frac{2\pi m k_B T}{h^2}\right)^{3/2} (V - V_{\text{ex}}) = \frac{\Delta V}{\Lambda^3}. \qquad (2.49)$$

The attractive mean field interaction between the clusters is considered in a similar way. According to Eq. 2.43, a common zero-of-energy reference has to be established first for the different cluster types $\{j\}$, which in most cases is set to the ground state energy of the relaxed (molecular) monomer unit. A physically reasonable energy scale for the ordering of the larger cluster structures is then obtained by considering the difference in the total ground state energy $E_{j,\text{tot}}$ of a cluster j and i_j-times the ground state energy of the relaxed monomer unit $E_{1,\text{tot}}$

$$\Delta E_{j,\text{intra}} = E_{j,\text{tot}} - i_j E_{1,\text{tot}}, \qquad (2.50)$$

where i_j again denotes the number of monomer units in cluster j. Energy differences of the form as in Eq. 2.50 are frequently encountered in quantum chemical applications and are referred to as *total/supramolecular interaction energies* or *interaction energies according to the supramolecular approach* [19]. Due to the limited basis sets normally applied in the actual quantum chemical calculation of these energies, a treatment of the basis set superposition error (bsse) is often reasonable, for instance via the counterpoise correction scheme [20]. In addition to this *intracluster* energy contribution, the van der Waals-like mean field interaction *between* the clusters is accounted for via a potential term of the form

$$\Delta E_{j,\text{inter}} = -\frac{a_{\text{mf}} i_j}{V}, \qquad (2.51)$$

where a_{mf} is an additional empirical parameter adjusting the strength of the intercluster interaction. As in the case of the classical van der Waals model, the factor i_j/V can be understood as a number density of the monomer units in cluster j per volume of the N-particle system. The sum of these two interaction energy terms is subsequently employed in the computation of the electronic cluster partition function $q_{j,\text{el}}$ according to [14, 15]

$$q_{j,\text{el}} = \exp\left[-\beta\left(\Delta E_{j,\text{intra}} + \Delta E_{j,\text{inter}}\right)\right] = \exp\left[-\beta\left(\Delta E_{j,\text{intra}} - \frac{a_{\text{mf}} i_j}{V}\right)\right], \qquad (2.52)$$

see also Eq. 2.42. From this derivation it is clear that in contrast to the conventional case in Eq. 2.42 true excited states of the monomer units or the clusters are not considered and that the applied energy "states" are referring to the binding situation within a cluster as well as the mean field interaction to the other clusters. Furthermore, it is apparent that the energy contributions in contrast to Eq. 2.43 are of a stabilizing origin, i.e., both the intracluster interaction energy (Eq. 2.50) as well as the intercluster interaction energy (Eq. 2.51) are negative, which results in electronic contributions to the overall cluster partition function larger than one. From a pure energetic point of view larger clusters exhibiting larger absolute values in both $\Delta E_{j,\mathrm{intra}}$ and $\Delta E_{j,\mathrm{inter}}$ are thus favored via the electronic contribution to the cluster partition function.

The presented van der Waals extension of the rrho model involves the introduction of two new parameters a_{mf} and b_{xv}, which can be understood as direct analogs to the van der Waals parameters a and b. There is no straightforward rule for the evaluation of these parameters on the basis of first principles methods, and the most convenient way of determining these unknowns is fitting some calculated thermodynamic quantity to an experimental reference. In this thesis the chosen reference quantity will normally be the molar volume, and the employed fitting procedure is a straightforward application of the commonly used least-squares fit criterion. According to that approach, a set of test isobars is computed over a predefined $a_{\mathrm{mf}}/b_{\mathrm{xv}}$ grid, and the absolute difference of each of these test curves to the (experimental) reference isobar is accumulated in an error vector of the form

$$\mathbf{\Delta V}(a_{\mathrm{mf}}, b_{\mathrm{xv}}) = \left[V_{\mathrm{ref}} - V_{a_{\mathrm{mf}},b_{\mathrm{xv}}}\right]_{T_{\mathrm{min}}\cdots T_{\mathrm{max}}}, \tag{2.53}$$

where $V_{\mathrm{ref}}, V_{a_{\mathrm{mf}},b_{\mathrm{xv}}}$ denote the molar reference volume and the volume of one of the test isobars at a given temperature, respectively, and the vector $\mathbf{\Delta V}$ has as many components as the number of sampled temperature points [17]. In a subsequent analysis the Euclidean norm $||\mathbf{\Delta V}||$ of each sampled vector $\mathbf{\Delta V}$ is computed, and the test isobar yielding the smallest norm is considered to be the most accurate approximation to the experimental reference. In the following, the Euclidean norm of the error vector $||\mathbf{\Delta V}||$ will also synonymously be denoted as the accuracy of the underlying test isobar [21, 22].

2.2.1.3 Cluster Equilibrium

The methodology presented so far covers the treatment of particle interactions on an accurate local scale (intracluster interactions via first principles computations) and an approximate non-local scale (intercluster interactions via a van der Waals-like mean field term) as well as the approximate treatment of the excluded volume in terms of atomic van der Waals radii. However, in order to obtain thermodynamic information for the N-particle system, the N-particle partition function (Eq. 2.46) has to be evaluated, which is only possible if the underlying cluster populations $\{n_j\}$ are known. In addition to the empirical suggestions made earlier, an algorithmic way for solving the population equations on the basis of a thermodynamic equilibrium between the different clusters has been developed in the late nineties of the last century [14].

The underlying equilibrium reaction takes into account the formation of all clusters in the set $\{C_j\}$ from the basic monomer unit C_1 according to [14, 15]

$$C_1 \leftrightarrows \frac{C_2}{2} \leftrightarrows \cdots \leftrightarrows \frac{C_j}{i_j} \leftrightarrows \cdots \leftrightarrows \frac{C_\eta}{i_\eta}, \tag{2.54}$$

where η denotes the largest cluster in the set and i_η equals the number of monomers in this cluster. If the equilibrium in Eq. 2.54 is assumed to be a thermodynamic equilibrium, the change in free energy A [1]

$$dA = -SdT - pdV + \sum_j \mu_j dn_j \tag{2.55}$$

has to be zero, where μ_j denotes the chemical potential of cluster j. At the conditions of the canonical ensemble (constant volume, constant temperature, constant particle number) this criterion is identical to the equality of the various chemical potentials $\{\mu_j\}$, i.e., Eq. 2.54 translates to

$$\mu_1 = \frac{\mu_2}{2} = \cdots = \frac{\mu_j}{i_j} = \cdots = \frac{\mu_\eta}{i_\eta}. \tag{2.56}$$

The chemical potential of a species j can most generally be expressed as a function of the canonical partition function Q according to [16]

$$\mu_j = -k_B T \left(\frac{\partial \ln(Q)}{\partial n_j} \right)_{n_{k \neq j}, V, T} \approx -k_B T \ln\left(\frac{q_j}{n_j}\right), \tag{2.57}$$

where the final transformation involves the application of Stirling's approximation [23]. The relation expressend in Eq. 2.57 provides a simple connection between the chemical potential μ_j and the population n_j of a given cluster j, and the substitution of the chemical potentials in Eq. 2.56 according to Eq. 2.57 yields a direct relation between the populations n_j, n_k of two different cluster species j and k. If one of these species (conveniently the monomer, $i_k = 1$) is set as a reference, the population of all remaining clusters can be expressed as functions of this reference population

$$\begin{aligned} n_j &= q_j \left[\frac{n_k}{q_k}\right]^{(i_j/i_k)} N_A^{(i_j-1)} \\ &= q_j \left[\frac{n_1}{q_1}\right]^{(i_j)} N_A^{(i_j-1)}. \end{aligned} \tag{2.58}$$

According to this approach, the problem of finding η populations is reduced to the determination of the reference population n_1. In order to obtain an equality for this reference population, the particle conservation condition in combination with a fixed total number of particles as expressed in Eq. 2.47 is employed. The substitution of the various populations in Eq. 2.47 according to Eq. 2.58 finally results in a polynomial equation in the monomer population n_1, which takes the form

$$0 = -N + \sum_{j=1}^{\eta} \left[\frac{(i_j) q_j N_A^{(i_j - 1)}}{q_1^{(i_j)}} \right] n_1^{(i_j)}. \tag{2.59}$$

Without loss of generality, the total number of particles N can be fixed to one mole, and according to the fundamental theorem of algebra there will be i_η different possible monomer population roots to Eq. 2.59 [24]. However, only roots from the real field are expected to be physically reasonable, and all complex solutions to Eq. 2.59 are discarded, but in the general case there will also be several real roots satisfying Eq. 2.59. Each of these roots can be employed to compute a complete cluster distribution on the basis of Eq. 2.58, which finally results in up to i_η possible cluster distributions ("phases") at the chosen state point. The stable phase at the applied conditions will be the one minimizing the free energy of the system, e.g., if the total particle number N, the temperature T, and the pressure p (instead of the volume V) are fixed, the related free energy will be the Gibbs free energy G, which can be obtained from the partition function Q according to[3]

$$G = -k_B T \ln(Q) + pV, \tag{2.60}$$

i.e., in order to evaluate the free energy, the volume of the different phases has to be obtained first. A straightforward way to this end exploits the relation between the (fixed) pressure p of the system and the canonical partition function $Q(N, V, T)$ [1, 14, 15]

$$p = kT \left(\frac{\partial \ln(Q)}{\partial V} \right)_{N,T}. \tag{2.61}$$

The contributions to the partition function Q which depend on the volume V are the translational partition function $q_{j,\mathrm{trans}}$ and the electronic partition function $q_{j,\mathrm{el}}$, see Eqs. 2.49 and 2.52. The evaluation of the derivative in Eq. 2.61 thus results in [16]

$$p = \frac{\sum_j^{\eta} \left(k_B T n_j V^2 - a_{\mathrm{mf}} i_j n_j \Delta V \right)}{V^2 \Delta V}, \tag{2.62}$$

[3] In principle one has to consider the isothermal–isobaric partition function $Q(N, T, p) = \sum_j \sum_V \exp(-\beta E_j) \exp(-\beta p V)$ for these thermodynamic reference variables, but since both summations are uncoupled, the pressure–volume term can be evaluated directly, yielding Eq. 2.60 [1].

where $\Delta V = V - V_{\mathrm{ex}}$ (see Eq. 2.49). Besides the volume V, this equation only contains known quantities and therefore can be rearranged to a polynomial equation in V taking the form

$$0 = [-p]V^3 + \left[k_{\mathrm{B}}T\sum_j^{\eta} n_j + pV_{\mathrm{ex}}\right]V^2 - \left[a_{\mathrm{mf}}\sum_j^{\eta} i_j n_j\right]V + \left[a_{\mathrm{mf}}\sum_j^{\eta} i_j n_j\right]V_{\mathrm{ex}}. \tag{2.63}$$

The coefficients of this polynomial equation depend on the cluster population distribution $\{n_j\}$, which means that for each of the different population sets up to three possible volumes can be obtained (the actual number is again equal to the number of real roots of Eq. 2.63). In combination with the population distributions, these volumes are employed for the determination of the stable phase at the chosen pressure p and temperature T according to Eq. 2.60, i.e., the population–volume combination which yields the minimum in G is considered to be the stable phase. The population set obtained in this way can subsequently be employed for the computation of the canonical partition function according to Eq. 2.46.

The above sketched scheme constitutes an algorithmic approach for the determination of the cluster distribution set and the phase volume of the stable phase at the chosen pressure–temperature conditions, which is readily implemented in a computer program [14, 16, 25]. However, in order to compute the coefficients of the population polynomial (Eq. 2.59), the cluster partition functions $\{q_j\}$ have to be available, but according to Eqs. 2.49 and 2.52 the translational and electronic contribution to each q_j depend on the volume V, which cannot be calculated until the populations are known, see Eq. 2.63. These mutual dependencies require an iterative procedure, in which a self-consistent set of cluster partition functions $\{q_j\}$, cluster populations $\{n_j\}$, and volume V is determined. The flow chart of the core iteration procedure implemented in the Peacemaker code is illustrated in Fig. 2.1 [16, 25]. The first step of the iteration consists of the computation of the cluster partition functions q_j, which depend on the phase volume V and also on the cluster populations $\{n_j\}$ via the excluded volume correction term, see Eq. 2.48. This means that in order to start the iteration, there are two initial guesses to be made, one for the volume V and one for the populations $\{n_j\}$. Depending on whether the current p, T state point is the first one of the calculation or an intermediate one, the initial guesses are either obtained by setting the volume V to the volume of an ideal gas at the chosen conditions and by distributing the number of monomer units C_1 uniformly over the different clusters according to Eq. 2.47 (i.e., $n_j = (i_j\eta)^{-1}$), or by constructing linear combinations of the volume and the populations obtained from previously converged iterations and the ideal gas volume and uniform distribution, respectively.

The next step includes the computation of the population polynomial coefficients, which according to Eq. 2.59 depend on the previously calculated cluster

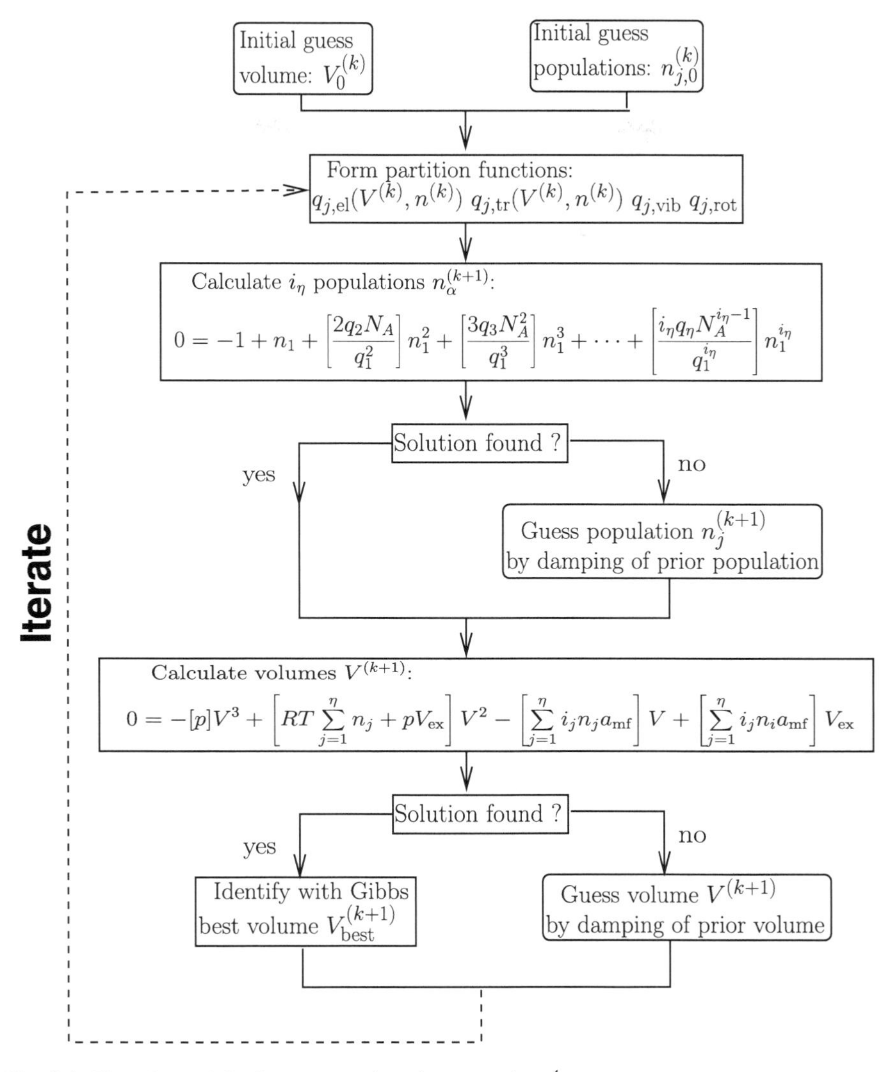

Fig. 2.1 Flowchart of the PEACEMAKER iteration procedure.[4]

partition functions. From the (real) roots of the population polynomial possible cluster population distributions $\{n_j\}$ are obtained (see Eq. 2.58), which are subsequently used for setting up the coefficients of the volume polynomial (Eq. 2.63). During the final step of each iteration, the Gibbs energy of all valid

[4] Reprinted with permission from Ref. [16]. Copyright 2005, American Institute of Physics

population–volume combinations is computed and the stable phase for the given p, T state point is identified. This population set and the corresponding volume reenter the cycle as the new guess for the next iteration. The quantity which is used to check the convergence of the iteration procedure is chosen to be the phase volume V in the PEACEMAKER code, but the populations $\{n_j\}$ or the partition function Q could be employed for this purpose as well [16, 25].

The algorithmic approach for the computation of the cluster populations of the van der Waals cluster gas introduced above is known as the *quantum cluster equilibrium* (*qce*) model in literature [14–16]. All relevant information required for a qce calculation can be obtained from static first principles calculations of the different cluster structures which are included in the equilibrium reaction, see Eq. 2.54.

2.2.2 Thermodynamics from Quantum Cluster Equilibrium Calculations

The quantum cluster equilibrium model introduced in the last section enables the computation of a self-consistent canonical partition function Q for the van der Waals cluster gas if a set of cluster structures (and corresponding properties like moments of inertia and harmonic frequencies) as well as a state point in the temperature–pressure space are specified as input values. Once the partition function is available, the computation of thermodynamic quantities is achieved in a straightforward manner, since all thermodynamic quantities can be expressed as analytical functions of the partition function Q. An example for such a relation has already been encountered in the derivation of the qce approach itself, namely the relation between the Gibbs free energy G (see Eq. 2.60) or the pressure p (see Eq. 2.61) and the partition function Q.

In all cases relevant for this thesis the functional dependancy between a quantity O and the partition function Q is of the form $O = f(\ln(Q))$, compare Eqs. 2.57, 2.60, and 2.61. This fact, combined with the particular form of the qce partition function (see Eq. 2.46) makes it possible to express the quantity O in terms of the cluster populations $\{n_j\}$ and the cluster partition functions $\{q_j\}$ instead of the N-particle partition function Q, thereby avoiding the computation of Q via a cumbersome evaluation of the $n_j!$ terms in Eq. 2.46 at all. Again, this observation has been employed before (see Eq. 2.57) and will be demonstrated in the following using the example of the entropy [17].

The functional dependancy between the entropy S and the canonical partition function Q is given by [1]

$$S = k_B \ln(Q) + k_B T \left(\frac{\partial \ln(Q)}{\partial T} \right)_{N,V}. \tag{2.64}$$

The replacement of Q in Eq. 2.64 in terms of the qce factorization from Eq. 2.46 results in

$$S = k_{\mathrm{B}} \sum_{j=1}^{\eta} \ln\left(\frac{q_j^{n_j}}{n_j!}\right) + k_{\mathrm{B}} T \left(\frac{\partial}{\partial T} \sum_{j=1}^{\eta} \ln\left(\frac{q_j^{n_j}}{n_j!}\right)\right). \tag{2.65}$$

In order to resolve the factorial terms in Eq. 2.65, Stirling's approximation ($\ln(n!) \approx n\ln(n) - n$) will be applied (as before in the case of the chemical potential, see Eq. 2.57), thereby yielding [23]

$$\begin{aligned} S &= k_{\mathrm{B}}\left[\sum_{j=1}^{\eta} n_j\left(\ln q_j - \ln n_j + 1\right)\right] + k_{\mathrm{B}} T\left[\frac{\partial}{\partial T}\sum_{j=1}^{\eta} n_j\left(\ln q_j - \ln n_j + 1\right)\right] \\ &= k_{\mathrm{B}}\left[\sum_{j=1}^{\eta} n_j\left(\ln q_j - \ln n_j + 1\right)\right] + k_{\mathrm{B}} T\left[\sum_{j=1}^{\eta} \frac{\partial}{\partial T} n_j \ln q_j\right]. \end{aligned} \tag{2.66}$$

The considerations presented in the last sections show that the cluster partition functions $\{q_j\}$ can be factorized into degree-of-freedom dependent contributions (see Eq. 2.44) and that these contributions are analytical functions of the temperature T (see Sect. 2.1.2), i.e., the evaluation of the temperature derivatives in the second term on the right hand side of Eq. 2.66 can be accomplished in a straightforward fashion according to

$$\begin{aligned} \left(\frac{\partial \ln q_j^{\mathrm{trans}}}{\partial T}\right)_{N,V} &= \frac{\partial}{\partial T}\ln\left[\frac{\Delta V(2\pi m_j k_{\mathrm{B}} T)^{3/2}}{h^3}\right] = \frac{3}{2T}; \\ \left(\frac{\partial \ln q_j^{\mathrm{rot}}}{\partial T}\right)_{N,V} &= \frac{\partial}{\partial T}\ln\left[\frac{\pi^{1/2}}{\sigma}\left(\frac{T^3}{\Theta_A\Theta_B\Theta_C}\right)^{1/2}\right] = \frac{3}{2T}; \\ \left(\frac{\partial \ln q_j^{\mathrm{vib}}}{\partial T}\right)_{N,V} &= \frac{\partial}{\partial T}\ln\left[\prod_{k}^{3M-6}\frac{\exp\left(-\frac{\beta h\nu_k}{2}\right)}{1-\exp(-\beta h\nu_k)}\right] \\ &= \sum_{k}^{3M-6}\frac{h\nu_k}{2k_{\mathrm{B}}T^2} + \frac{\frac{h\nu_k}{k_{\mathrm{B}}T^2}}{\exp(\beta h\nu_k)-1}; \\ \left(\frac{\partial \ln q_j^{\mathrm{el}}}{\partial T}\right)_{N,V} &= \frac{\partial}{\partial T}\ln\left[\exp\left(\frac{-\Delta E_{j,\mathrm{intra}} + i_j a_{\mathrm{mf}} V^{-1}}{k_{\mathrm{B}} T}\right)\right] \\ &= -\frac{-\Delta E_{j,\mathrm{intra}} + i_j a_{\mathrm{mf}} V^{-1}}{k_{\mathrm{B}} T^2}. \end{aligned} \tag{2.67}$$

The insertion of the degree-of-freedom dependent temperature derivatives from Eq. 2.67 in Eq. 2.66 leads to the final qce entropy expression [17]

$$S = k_{\mathrm{B}} \sum_{j}^{\eta} n_j (\ln q_j - \ln n_j + 1)$$
$$+ k_{\mathrm{B}} T \sum_{j}^{\eta} n_j \left[\frac{\Delta E_{j,\mathrm{intra}} - i_j a_{\mathrm{mf}} V^{-1}}{k_{\mathrm{B}} T^2} + \left(\sum_{k}^{3M-6} \frac{h\nu_k}{2k_{\mathrm{B}} T^2} + \frac{\frac{h\nu_k}{k_{\mathrm{B}} T^2}}{\exp\left(\frac{h\nu_k}{k_{\mathrm{B}} T}\right) - 1} \right) + \frac{3}{T} \right]. \tag{2.68}$$

The necessary quantities which have to be available for the entropy calculation according to Eq. 2.68 are the cluster populations $\{n_j\}$, the cluster partition functions $\{q_j\}$, and the cluster properties which are also needed to calculate the set $\{q_j\}$. As a second example, the equality for the qce isochoric heat capacity is given, which can be computed according to [26]

$$C_{\mathrm{v}} = \left(\frac{\partial U}{\partial T} \right)_{N,V} = 2k_{\mathrm{B}} T \left(\frac{\partial \ln(Q)}{\partial T} \right)_{N,V} + k_{\mathrm{B}} T^2 \left(\frac{\partial^2 \ln(Q)}{\partial T^2} \right)_{N,V}$$
$$= 2k_{\mathrm{B}} T \sum_{j}^{\eta} n_j \frac{\partial}{\partial T} \ln(q_j) + k_{\mathrm{B}} T^2 \sum_{j}^{\eta} n_j \frac{\partial^2}{\partial T^2} \ln(q_j), \tag{2.69}$$

where the temperature derivatives of the cluster partition functions can be evaluated in an analytical way as in the case of the entropy, see Eq. 2.67. In complete analogy to the approach which led to Eq. 2.68, analytical equations for all the other relevant thermodynamic quantities can be derived on the basis of the qce partition function in Eq. 2.46. Although the qce partition function is based on a non-interacting van der Waals cluster gas and thereby is of a profoundly approximate nature concerning the treatment of real condensed systems, the calculation of thermodynamic properties in the frame of the qce approach is realized in an analytically exact way (with the exception of Stirling's approximation), which can be considered to be one of the most important advantages of this method. The availability of an analytical partition function is not given in most other approaches suitable for the treatment of condensed phase thermodynamics, and in many cases such approaches have to rely on approximate relations for the relevant thermodynamic quantities depending on additional parameters or on the computational details, e.g., the simulation length [27–29]. Some of these approaches will be outlined briefly in the next section.

2.3 Thermodynamic Data from Molecular Dynamics Simulations

2.3.1 Equilibrium Methods

In contrast to the static approaches discussed in the preceeding sections, molecular dynamics (md) simulations explicitly account for the microscopic dynamics of the investigated system, i.e., each md simulation evolves on a real time axis. In the case of an equilibrium md simulation, thermodynamic constraints are placed on the system according to the thermodynamic nature of the process being investigated, for instance one would fix the number of particles N, the volume V, and the temperature T in the case of an isochoric system being in equilibrium with a heat bath. A md simulation in this setup is equivalent to sampling the canonical (N, V, T) ensemble, for which the relevant thermodynamic potential is the Helmholtz free energy A given by

$$A = -k_{\mathrm{B}}T\ln(Q(N,V,T)). \tag{2.70}$$

As outlined above, both the rrho as well as the qce approach are fundamentally linked to the factorization of the N-particle partition function Q into single particle contributions, thereby ignoring all interparticle interactions as a first approximation. However, md simulation methods explicitly incorporate interparticle interactions via an interaction potential $U(\mathbf{q})$, which is either obtained from electronic structure calculations on the fly (*first principles* molecular dynamics (fpmd) simulations) [30, 31] or from an analytical force field function (traditional molecular dynamics simulations) [32]. Thus, an ideal-gas-like factorization approach for the determination of Q cannot be brought in line with the basic methodology of md simulations. On the other hand, a direct evaluation of the N-particle partition function is prohibitively difficult, which demonstrates the need for a fundamentally different approach.

In most cases, it is not necessary to calculate the absolute value for, e.g., A according to Eq. 2.70 in order to extract the relevant thermodynamic information, since most chemical processes involve a change from a well-defined initial state 1 to a final state 2, and it is often sufficient to have access to the change of a thermodynamic quantity during the reaction course $1 \rightarrow 2$. If the quantity of interest is the free energy change ΔA, Eq. 2.70 reduces to

$$\Delta A = A_2 - A_1 = -k_{\mathrm{B}}T\ln\left(\frac{Q_2}{Q_1}\right), \tag{2.71}$$

and if the masses of the particles do not change during the process $1 \rightarrow 2$, Eq. 2.5 can be employed to obtain

$$\Delta A = -k_{\mathrm{B}}T\ln\left(\frac{Z_2}{Z_1}\right). \tag{2.72}$$

This equation indicates that the new target quantity is the ratio Z_2/Z_1, which often is still too complicated to calculate for systems with many degrees of freedom if no further treatment is applied, and the largest part of this section will be devoted to transformations of Eqs. 2.71 and 2.72, respectively, to expressions which can readily be combined with the methodology of md simulations. The two emerging methods covered here are referred to as free energy perturbation (fep) theory and thermodynamic integration, which are among the most frequently employed approaches for the calculation of free energy changes from md simulations today [33, 34].

One of the most powerful and universal approaches of applied mathematics is perturbation theory, which divides a problem too complicated to be exactly solvable into a (simpler) reference problem and a perturbation. In the frame of the Hamiltonian formalism introduced in Sect. 2.1.1 such a division is often realized in the following way [33]

$$\mathcal{H}_{\mathrm{tot}}(\mathbf{p},\mathbf{q}) = \mathcal{H}_{\mathrm{ref}}(\mathbf{p},\mathbf{q}) + \Delta\mathcal{H}(\mathbf{p},\mathbf{q}). \tag{2.73}$$

In this equation $\mathcal{H}_{\mathrm{tot}}(\mathbf{p},\mathbf{q})$ refers to the Hamiltonian of the full problem (compare also Eq. 2.2), and $\mathcal{H}_{\mathrm{ref}}(\mathbf{p},\mathbf{q}), \Delta\mathcal{H}(\mathbf{p},\mathbf{q})$ denote the Hamiltonian of the simpler reference problem and of the perturbation, respectively. For instance, the full problem could be a solute at infinite dilution in a solvent. In this case the reference problem $\mathcal{H}_{\mathrm{ref}}$ could be chosen to be the unsolvated solute, and the perturbation $\Delta\mathcal{H}$ would consist of all the solute–solvent interactions[5] [33]. Following Eqs. 2.1 and 2.71, the free energy change between the perturbed system 2 ($\mathcal{H}_2 = \mathcal{H}_{\mathrm{tot}}$) and the reference system 1 ($\mathcal{H}_1 = \mathcal{H}_{\mathrm{ref}}$) can be expressed as

$$\begin{aligned}\Delta A &= -k_{\mathrm{B}}T\ln\left[\frac{\int\exp(-\beta\mathcal{H}_2(\mathbf{p},\mathbf{q}))d\mathbf{p}d\mathbf{q}}{\int\exp(-\beta\mathcal{H}_1(\mathbf{p},\mathbf{q}))d\mathbf{p}d\mathbf{q}}\right]\\ &= -k_{\mathrm{B}}T\ln\left[\frac{\int\exp(-\beta\Delta\mathcal{H}(\mathbf{p},\mathbf{q})\exp(-\beta\mathcal{H}_{\mathrm{ref}}(\mathbf{p},\mathbf{q}))d\mathbf{p}d\mathbf{q}}{\int\exp(-\beta\mathcal{H}_{\mathrm{ref}}(\mathbf{p},\mathbf{q}))d\mathbf{p}d\mathbf{q}}\right].\end{aligned} \tag{2.74}$$

In the frame of the (N, V, T) ensemble the probability P for the system under investigation to be in a certain energy state E_j (or to occupy a certain region of phase space $d\mathbf{p}d\mathbf{q}$ in a classical treatment) is proportional to the corresponding Boltzmann factor $\exp(-\beta E_j)$, and the proportionality constant is the inverse of the partition function Q, which ensures a normalization of the overall probability to one [6]. According to that reasoning, the probability distribution function $P_1(\mathbf{p},\mathbf{q})$ for the reference state 1 is given as

$$P_1(\mathbf{p},\mathbf{q}) = \frac{\exp(-\beta\mathcal{H}_1(\mathbf{p},\mathbf{q}))}{Q(N,V,T)} \propto \frac{\exp(-\beta\mathcal{H}_1(\mathbf{p},\mathbf{q}))}{\int\exp(-\beta\mathcal{H}_1(\mathbf{p},\mathbf{q}))d\mathbf{p}d\mathbf{q}}. \tag{2.75}$$

With this definition, the free energy change for the process $1 \to 2$ can be formulated in terms of the probability distribution function [33]

$$\begin{aligned}\Delta A &= -k_{\mathrm{B}}T\ln\left[\int\exp(-\beta\Delta\mathcal{H}(\mathbf{p},\mathbf{q}))P_1(\mathbf{p},\mathbf{q})d\mathbf{p}d\mathbf{q}\right]\\ &= -k_{\mathrm{B}}T\ln[\langle\exp(-\beta\Delta\mathcal{H}(\mathbf{p},\mathbf{q}))\rangle_1],\end{aligned} \tag{2.76}$$

where the brackets $\langle\ldots\rangle$ indicate a so-called *ensemble average* (and no scalar product as in the case of Sect. 2.1.1). The direct computation of the ensemble average would require the availability of the probability distribution function

---

[5] Depending on the exact nature of the problem and the quantity of interest, the solvent-solvent interactions would have to be included in the perturbation term as well.

$P_1(\mathbf{p}, \mathbf{q})$ and thereby of the partition function Q (see Eq. 2.75), but under the assumption of the ergodic hypothesis an md simulation in the (N, V, T) ensemble samples the phase space of the examined system just according to the probability distribution in Eq. 2.75 and thus intrinsically provides the correct statistical weight for the average in Eq. 2.76 [29]. It should be noted that in cases where the particle masses do not change during the transformation $1 \rightarrow 2$, an identical derivation starting from Eq. 2.72 yields the equality

$$\Delta A = -k_B T \ln[\langle \exp(-\beta \Delta U(\mathbf{q})) \rangle_1], \tag{2.77}$$

where $\Delta U(\mathbf{q})$ denotes the perturbation in the potential in analogy to Eq. 2.73. The result expressed in Eqs. 2.76 and 2.77 demonstrates that in order to obtain the free energy change for the process $1 \rightarrow 2$ it is only necessary to sample the reference system 1 and to employ the obtained trajectory for the evaluation of the exponential average in Eq. 2.76. Despite this simple procedure, there might be practical problems in the evaluation of Eq. 2.76. In order to illustrate these aspects, it is helpful to rewrite Eq. 2.76 in terms of the probability distribution function of the perturbation $P(\Delta U)$, which thereby reduces to a one-dimensional integral according to [33]

$$\Delta A = -k_B T \ln\left[\int \exp(-\beta \Delta U) P_1(\Delta U) d\Delta U\right]. \tag{2.78}$$

In many cases, $P_1(\Delta U)$ will be roughly of Gaussian-like shape. However, the integrand being relevant for the free energy change is not $P_1(\Delta U)$ but the product between $P_1(\Delta U)$ and $\exp(-\beta \Delta U)$, which is also of Gaussian-like form but shifted to lower ΔU. This means that the most important contributions to the integral in Eq. 2.78 will occur with a relatively low probability in the sampling of $P_1(\Delta U)$ if the two distributions do not overlap significantly, thereby limiting the general applicability of Eqs. 2.76 and 2.77, respectively [33]. Another problem arises if the initial and final states of the system under investigation are very different in a sense that their important regions in phase space either only overlap partly or do not overlap at all. In these cases the perturbation $\Delta \mathcal{H}$ will be rather large, and an enhanced sampling procedure has to be applied. A straightforward approach considers the construction of intermediate states $\{i\}$ representing only sections of the overall process $1 \rightarrow 2$. Each of these intermediate processes can then be treated separately in terms of the fep approach, and the total change in free energy is obtained as

$$\Delta A = \sum_i \Delta A_{i,i+1} = -k_B T \ln\left[\langle \exp(-\beta \Delta U_{i,i+1}) \rangle_i\right]. \tag{2.79}$$

The second method for the calculation of free energy changes on the basis of md simulations presented here is the thermodynamic integration scheme [29, 34]. The basic idea behind this approach can be summarized as instead of calculating the free energy directly, e.g., via Eq. 2.70, the derivative of the free energy with respect to an (arbitrary) order parameter λ is computed. This is a most natural approach which in a similar way is also employed in real experiments, where only relative changes of the free energy with respect to an external parameter (e.g. the volume) are measureable [29]. The order parameter λ can either be a function of the (generalized) coordinates of the system (e.g. the distance between two particles) or be a parameter in the Hamiltonian $\mathcal{H}$, which is used to switch a certain interaction on or off [34]. This latter variant is also referred to as an alchemical transformation (since the system can undergo significant changes during the process $1 \rightarrow 2$, see below) and it will be the one which is treated here exemplarily. The parametrization of the Hamiltonian is usually carried out in the following way

$$\mathcal{H}_\lambda(\mathbf{p},\mathbf{q}) = (1-\lambda)\mathcal{H}_1(\mathbf{p},\mathbf{q}) + \lambda\mathcal{H}_2(\mathbf{p},\mathbf{q}), \tag{2.80}$$

so that for $\lambda = 0$ the initial state 1 and for $\lambda = 1$ the final state 2 is obtained. The free energy of the system under investigation can readily be expressed in terms of this parameterized Hamiltonian via the partition function from Eq. 2.1. The derivative of this free energy expression with respect to the order parameter λ is given as

$$\frac{dA(\lambda)}{d\lambda} = \frac{\int\left(\frac{\partial\mathcal{H}_\lambda(\mathbf{p},\mathbf{q})}{\partial\lambda}\right)\exp(-\beta\mathcal{H}_\lambda(\mathbf{p},\mathbf{q}))d\mathbf{p}d\mathbf{q}}{\int\exp(-\beta\mathcal{H}_\lambda(\mathbf{p},\mathbf{q}))d\mathbf{p}d\mathbf{q}}. \tag{2.81}$$

This equation is again of the form $\int O(\lambda)P_\lambda\, d\mathbf{p}d\mathbf{q}$ (compare Eq. 2.75) and thus can be identified as an ensemble average for $O = \partial\mathcal{H}_\lambda/\partial\lambda$. Thus Eq. 2.81 can be rewritten as

$$\frac{dA(\lambda)}{d\lambda} = \left\langle\frac{\partial\mathcal{H}_\lambda(\mathbf{p},\mathbf{q})}{\partial\lambda}\right\rangle_\lambda, \tag{2.82}$$

and the free energy change for the process $1 \rightarrow 2$ follows from integrating Eq. 2.82 along λ according to

$$\Delta A = \int\limits_{\lambda=0}^{\lambda=1}\frac{dA(\lambda)}{d\lambda}d\lambda = \int\limits_{\lambda=0}^{\lambda=1}\left\langle\frac{\partial\mathcal{H}_\lambda(\mathbf{p},\mathbf{q})}{\partial\lambda}\right\rangle_\lambda d\lambda. \tag{2.83}$$

The free energy change ΔA can thus be computed from ensemble averages of the change in the Hamiltonian under variation of λ, which are typically obtained from a series of equilibrium md simulations carried out at different values for λ between 0 and 1 [29]. This approach is a very general one and will be applicable as long as $\mathcal{H}_\lambda$ interpolates smoothly between $\mathcal{H}_1$ and $\mathcal{H}_2$. The path along which the order parameter is varied not necessarily has to be a physical one. For instance, it is possible to calculate the difference in free energy upon transformation of a functional group within a molecule into another one by gradually changing the first group into the second at different values of λ [29]. It is clear that during such a transformation the particle masses will possibly change, which is the reason for parameterizing the full Hamiltonian (see Eq. 2.80) and not only the configurational contribution U. If, however, the particle masses stay constant during the process $1 \rightarrow 2$, Eq. 2.83 again simplifies to

$$\Delta A = \int_{\lambda=0}^{\lambda=1} \left\langle \frac{\partial U_\lambda(\mathbf{q})}{\partial \lambda} \right\rangle_\lambda d\lambda. \tag{2.84}$$

It should be noted that additional weighting factors can be included to each of the ensemble averages used to compute the free energy change according to Eq. 2.83, which help to stress the influence of certain λ-stages on the path $\lambda = 0 \rightarrow \lambda = 1$ [34].

The discussion so far concentrated on the calculation of free energy changes in the (N, V, T) ensemble, and due to the missing availability of the partition function in md simulations, there is no general approach for the computation of other thermodynamic quantities like e.g. the entropy. However, the methodologies of perturbation theory and thermodynamic integration are often applicable to other properties in a similar way. For instance, entropy differences in the (N, V, T) ensemble can be expressed in the thermodynamic integration scheme in a straightforward way following the same route as in the case of free energy changes [35]

$$\Delta S = S_2 - S_1 = \frac{1}{k_B T^2} \int_{\lambda=0}^{\lambda=1} \left[\left\langle \frac{\partial \mathcal{H}(\lambda)}{\partial \lambda} \right\rangle_\lambda \langle \mathcal{H}(\lambda) \rangle_\lambda - \left\langle \frac{\partial \mathcal{H}(\lambda)}{\partial \lambda} \mathcal{H}(\lambda) \right\rangle_\lambda \right]. \tag{2.85}$$

Another approach is based on the application of classical thermodynamic relations, e.g. between the free energy A and the entropy S

$$S = -\left(\frac{\partial A}{\partial T} \right)_{N,V}, \tag{2.86}$$

which can be evaluated with the aid of Eqs. 2.70 and 2.1 according to [35]

$$\begin{aligned} S &= k_B \ln(Q) + \frac{\int \mathcal{H}(\mathbf{p},\mathbf{q}) \exp(-\beta \mathcal{H}(\mathbf{p},\mathbf{q})) d\mathbf{p} d\mathbf{q}}{N! h^{3N} Q T} \\ &= k_B \ln(Q) + \frac{\langle \mathcal{H} \rangle}{T} = \frac{-A + \langle \mathcal{H} \rangle}{T}; \\ \Delta S &= \frac{\Delta \langle \mathcal{H} \rangle - \Delta A}{T}. \end{aligned} \tag{2.87}$$

The changes in free energy ΔA are either obtained via the fep scheme or the thermodynamic integration method, and the energy change $\Delta \langle \mathcal{H} \rangle$ can be estimated from the difference in total energy between the two considered states 1 and 2 [35]. Finally, the derivative in Eq. 2.86 could also be evaluated by numerical methods, for instance in terms of the finite difference method, which yields approximate equations of the form [33, 35]

$$\Delta S = -\frac{\Delta A(T + \Delta T) - \Delta A(T - \Delta T)}{2\Delta T}, \tag{2.88}$$

where ΔT is an appropriate temperature interval and $\Delta A(T + \Delta T), \Delta A(T - \Delta T)$ denote the free energy change for the process $1 \rightarrow 2$ at the temperatures $T + \Delta T$ and $T - \Delta T$, respectively. Is is clear from Eq. 2.88 that this approach relies on the approximation of a linear dependancy between free energy and temperature in the interval $2\Delta T$.

Besides the entropy, thermodynamic response functions as, e.g., heat capacities often yield important contributions to the thermodynamic characterization of a system. In the frame of md simulations, such response functions are most generally expressed as functions of the ensemble averages of fluctuations $\langle \delta O^2 \rangle$ in a mechanical quantity O, which are defined as [32]

$$\langle \delta O^2 \rangle = \langle O^2 \rangle - \langle O \rangle^2. \tag{2.89}$$

In the case of the (N, V, T) ensemble the relevant specific heat capacity is the isochoric heat capacity C_V related to fluctuations in the Hamiltonian according to [32]

$$C_V = \frac{\langle \delta \mathcal{H}^2 \rangle}{k_B T^2}. \tag{2.90}$$

It should be noted that in general estimates of changes in the free energy components ΔH and ΔS via the above sketched approaches are more inaccurate than the calculation of the corresponding free energy change via fep or thermodynamic integration, which is the reason why free energy changes have been discussed in greater detail in this section [33, 36]. Suggestions have been made that a careful sampling protocol (e.g., the consideration of additional sampling stages as in the case of the reasoning behind Eq. 2.79) is of higher importance for accurate estimates of ΔH and ΔS than the actual method of computation [36]. In addition, system specific information might be used to guide the choice of an appropriate computational method, e.g., the calculation of entropy changes in a system near a phase transition via Eq. 2.88 could give rise to problems [33].

2.3.2 A Unified Nonequilibrium Approach

The considerations of the previous part concentrated on two important methods for the computation of free energy changes for a process $1 \rightarrow 2$ on the basis of md simulations, namely free energy perturbation (fep) and thermodynamic integration. These two approaches can be understood as the limiting cases of a more general framework which will be covered in the present section.

The basic idea behind the fep scheme is the sampling of an equilibrium distribution at the initial state 1, from which the *instantaneous* energy differences to the final state 2 are calculated and exponentially weighted, see Eqs. 2.76 and 2.77. From a formal point of view, there is no sampling of intermediate states or step-by-step changing of the order parameter λ, although the consideration of intermediate stages can considerably contribute to the accuracy of the computed free energy change (compare Eq. 2.79). In contrast, the thermodynamic integration approach relies on the generation of an equilibrium distribution at each individual λ stage, and the free energy change is computed along this equilibrium path from $\lambda = 0$ to $\lambda = 1$ by summing up (integrating) the changes in the Hamiltonian resulting from the variation in λ, see Eqs. 2.83 and 2.84. Along this path, the system is always in equilibrium, which means that the whole process is completely reversible and no dissipation of work into heat takes place. Thus, the work W necessary to initiate (or gained from) the process $1 \rightarrow 2$ directly corresponds to the change in free energy. In a more general framework, one could express the change from $\lambda = 0$ to $\lambda = 1$ as a continuous, time-dependent process, thereby introducing a time dependancy in the order parameter $\lambda \rightarrow \lambda(t)$ and thus in the Hamiltonian. If the rate at which λ changes is again very slow, the system will stay in (or near the) equilibrium during the whole process, and the thermodynamic integration formula (see Eq. 2.83) can be rewritten as [37]

$$\Delta A = \lim_{\tau \rightarrow \infty} \int_0^\tau \left. \frac{\partial \mathcal{H}}{\partial \lambda} \right|_{\lambda=\lambda(t)} \dot{\lambda}(t) dt, \tag{2.91}$$

where the dot denotes the rate at which $\lambda(t)$ changes, i.e., the time derivative of λ, and the limit $\tau \rightarrow \infty$ indicates an infinitely slow changing rate, thereby ensuring reversibility. If in contrast the transformation from $\lambda = 0$ to $\lambda = 1$ is done at a *finite* rate, the process will no longer be reversible, since the system will be driven out of equilibrium as the process proceeds. During this irreversible change, the necessary (or gained) work W no longer corresponds to the change in free energy due to the dissipation of work into heat, which implies that on average the performed work is larger than the free energy change

$$\langle W(\tau) \rangle \geq \Delta A, \tag{2.92}$$

where the equality in Eq. 2.92 only holds for the limiting case of a reversible change. Equation 2.92 can be understood as a formulation of the second law of thermodynamics in terms of work and dissipation. Following this reasoning, Eq. 2.91 for finite τ is given as [37]

$$W(\tau) = \int_0^\tau \left. \frac{\partial \mathcal{H}}{\partial \lambda} \right|_{\lambda=\lambda(t)} \dot{\lambda}(t) dt. \tag{2.93}$$

So far, the treatment of a time-dependent change in the order parameter $\lambda(t)$ only led to an inequality between work (which can be obtained from nonequilibrium md simulations) and the change in free energy, see Eq. 2.92. However, Jarzynski demonstrated that the inequality in Eq. 2.92 can be transformed into an equality if instead of W and ΔA the corresponding exponentially weighted quantities are considered [37, 38]

$$\langle \exp(-\beta W(\tau)) \rangle = \exp(-\beta \Delta A). \tag{2.94}$$

Equation 2.94 is widely known as the Jarzynski equality and constitutes a more general framework of the equilibrium concepts which underlie the fep and thermodynamic integration scheme, namely the determination of the free energy change ΔA of a process $1 \rightarrow 2$ for changes in the Hamiltonian (via $\lambda(t)$) at an *arbitrary* rate. There exist several ways for the derivation of Eq. 2.94 (which will not be covered here), of which the relation between the equilibrium Boltzmann distribution and path integral ensemble averages according to the Feynman-Kac theorem is possibly the most instructive and useful variant [37, 39–41].

There are some caveats concerning the practical application of Eq. 2.94 to be considered, though. First of all, the integration of the equations of motion (in case of traditional md simulations) for a time-dependent Hamiltonian can generally lead to larger errors in the trajectories using the conventional integration schemes [32, 37]. Furthermore, the equilibrium distribution for $\lambda = 0$ from which the nonequilibrium trajectories will be generated has to be sampled sufficiently in a way that the different initial configurations are uncorrelated. If this is the case, the corresponding work averages in Eq. 2.94 will also be uncorrelated, thus enabling the application of advanced statistical methods for the free energy estimate like e.g. Bennett's acceptance ratio approach, which additionally considers the backward direction path (i.e., going from $\lambda = 1$ to $\lambda = 0$) [42, 43]. Such approaches are especially important in the light of the sampling problems discussed for the fep approach in the last section. The comparison of Eqs. 2.94 and 2.78 shows that in both cases the ensemble average is taken over the exponential of a quantity and not over the quantity itself, and as pointed out in the discussion of Eq. 2.78 important contributions to the ensemble average might occur with relatively low probability if both probability distributions do not overlap significantly. In addition, the probability distribution of the work values strongly depends on the transformation path chosen for the process $1 \rightarrow 2$. The occurrence of large barriers which are crossed under tension or a transformation carried out too rapidly without sampling relevant parts of the corresponding phase space can lead to broad work distributions and an increased effect of dissipation [37]. In such cases the obtained free energy change can be subject to a systematical error which will not disappear even if the statistical error is reduced. A possible way to obtain work distributions of smaller width lies in the consideration of additional stages in the overall process $1 \rightarrow 2$ as introduced for the fep methodology in the last section, see Eq. 2.79. However, this approach significantly increases the computational effort, since in order to obtain starting points for the nonequilibrium trajectories a new equilibrium distribution has to be sampled for each new λ stage [37].

References

1. McQuarrie DA (1976) Statistical mechanics. Harper and Row, New York
2. Kirkwood JG (1933) Phys Rev 44:31–37

3. Kirkwood JG (1934) Phys Rev 45:116–117
4. Jensen F (1999) Introduction to computational chemistry. Wiley-VCH, Chichester
5. Levine IN (2000) Quantum Chemistry. Prentice Hall, New Jersey
6. Pathria RK (1996) Statistical mechanics, 2nd edn. Elsevier, Oxford
7. ter Haar D (1995) Elements of statistical mechanics. Butterworth–Heinemann Ltd, Oxford
8. Reinhold J (2004) Quantentheorie der Moleküle. Teubner, Wiesbaden
9. Szabo A, Ostlund NS (1996) Modern quantum Chemistry. Dover, New York
10. Chandler D (1987) Introduction to modern statistical mechanics. Oxford University Press, New York
11. McQuarrie DA, Simon JD (1997) Physical Chemistry. University Science Books, Sausalito
12. Ahlrichs R, Bär M, Häser M, Horn H, Kölmel C (1989) Chem Phys Lett 162:165–169
13. Neugebauer J, Reiher M, Kind C, Hess BA (2002) J Comput Chem 23:895–910
14. Weinhold F (1998) J Chem Phys 109:367–372
15. Kirchner B (2007) Phys Rep 440:1–111
16. Kirchner B (2005) J Chem Phys 123:204116
17. Spickermann C, Lehmann SBC, Kirchner B (2008) J Chem Phys 128:244506
18. Silla E, Tunon I, Pascual-Ahuir JL (1991) J Comput Chem 12:1077–1088
19. Ponti A, Mella M (2003) J Phys Chem A 107:7589–7596
20. Boys SF, Bernardi F (1970) Mol Phys 19:553–566
21. Lehmann SBC, Spickermann C, Kirchner B (2009) J Chem Theory Comput 5:1640–1649
22. Lehmann SBC, Spickermann C, Kirchner B (2009) J Chem Theory Comput 5:1650–1656
23. Bronstein IN, Semendjajew KA, Musiol G, Mühlig H (2006) Taschenbuch der Mathematik. Verlag Harri Deutsch, Frankfurt am Main
24. Smithies F (2000) Notes Rec R Soc 54:333–341
25. PEACEMAKER V 1.4 Copyright B. Kirchner, written by Kirchner B, Spickermann C, Lehmann SBC, Perlt E, Uhlig F, Langner J, Domaros Mv and Reuther p 2004–2009, University of Bonn, Institute of Physical and Theoretical Chemistry, University of Leipzig, Wilhelm-Ostwald Institute of Physical and Theoretical Chemistry Bonn-Leipzig 2009, see also http://www.uni-leipzig.de/~quant/index.html/
26. Perlt E (2008) Isochore Wärmekapazität und Quantum Cluster Equilibrium-Methode, Thesis, Universität Leipzig
27. Fredenslund A, Jones RL, Prausnitz JM (1975) AIChE J 21:1086–1099
28. Fredenslund A, Gmehling J, Rasmussen P (1977) Vapor-liquid equilibria using UNIFAC-A group contribution method. Elsevier, Amsterdam
29. Frenkel D, Smit B (2002) Understanding molecular simulations. Academic press, San Diego
30. Car R, Parrinello M (1985) Phys Rev Lett 55:2471–2474
31. Marx D, Hutter J (2009) Ab initio molecular dynamics: basic theory and advanced methods. Cambridge University Press, Cambridge
32. Allen MP, Tildesley DJ (1987) Computer simulations of liquids. Claredon Press, Oxford
33. Chipot C, Pohorille A (2007) Calculating free energy differences using perturbation theory. In: Chipot C, Pohorille A (eds) Free energy calculations. Springer, Berlin
34. Darve E (2007) Thermodynamic integration using constrained and unconstrained dynamics. In: Chipot C, Pohorille A (eds) Free energy calculations. Springer, Berlin
35. Peter C, Oostenbrink C, van Dorp A, van Gunsteren WF (2004) J Chem Phys 120:2652–2661
36. Lu N, Kofke DA, Woolf TB (2003) J Phys Chem B 107:5598–5611
37. Hummer G (2007) Nonequilibrium methods for equilibrium free energy calculations. In: Chipot C, Pohorille A (eds) Free energy calculations. Springer, Berlin
38. Jarzynski C (1997) Phys Rev Lett 78:2690–2693
39. Crooks GA (2000) Phys Rev E 61:2361–2366
40. Kac M (1949) Trans Am Math Soc 65:1–13
41. Hummer G, Szabo A (2005) Acc Chem Res 38:504–513
42. Bennett CH (1976) J Comput Phys 22:245–268
43. Cossins BP, Foucher S, Edge CM, Essex JW (2009) J Phys Chem B 113:5508–5513

Chapter 3
Assessment of the Rigid Rotor Harmonic Oscillator Model at Increased Densities

Following the methodology introduced in the preceding chapter, the most straightforward approach for the calculation of thermodynamic quantities from atomistic calculations, namely the rigid rotor harmonic oscillator model, will be tested and evaluated in the following. The rrho model is explicitly formulated for the conditions of an ideal gas and thereby per definition is only a gross approximation to the condensed state of matter. Nevertheless, this approach is employed frequently for the prediction of thermochemical reaction data in combination with static first principles computations also for reactions taking place in solution. In this chapter, supramolecular architectures of the rotaxane type will be applied as test systems of high chemical complexity, thereby providing a demanding challenge for the model. The second part concentrates on a detailed analysis of the errors and problems occurring in this simplified approach and helps to set the stage for the application of more elaborate models.[1]

3.1 Supramolecular Compounds as Test Systems of High Complexity

3.1.1 The Rotaxane Architecture

The term "rotaxane" refers to a certain class of supramolecular architectures in which an axle-like molecular structure is combined with a macrocyclic compound in such a way that a threaded arrangement is created [4]. In this arrangement, the axle is held in place inside the cavity of the wheel in terms of a *mechanical bond*. Such a bond does not constitute a chemical bond in the usual sense, but nevertheless makes the breaking of a chemical bond necessary in order to separate the

[1] Parts of this chapter have already been published in [1–3].

C. Spickermann, *Entropies of Condensed Phases and Complex Systems*, Springer Theses, DOI: 10.1007/978-3-642-15736-3_3,

two components. The mechanical bond in the rotaxane architecture is normally realized in terms of sterically demanding substituents at both ends of the axle (so-called *stoppers*), which prevent the wheel from dethreading off the axle [4]. In rotaxane syntheses, the threaded arrangement between wheel and axle is typically established first by exploiting some form of the so-called *template effect*, which preorganizes the threaded arrangement in terms of non-covalent interactions, and the stoppers are added during the final step, thereby fixing the preorganized structure [5–7]. The preorganized intermediates without stoppers are commonly referred to as *pseudorotaxanes*, because they already show the final spatial arrangement of wheel and axle, but the mechanical bond is not yet established and the stability of the threaded arrangement solely depends on the strength of the template effect between both components [8]. A typical example of a pseudorotaxane complex is illustrated in Fig. 3.1.

The ball and stick model shown in Fig. 3.1 indicates two wheel-to-axle amide hydrogen bonds labeled as "wa_1" and "wa_2" as well as a single axle-to-wheel hydrogen bond labeled as "aw". The orientation of the four amide groups of the macrocycle is also indicated via the labels "in" and "out", where "out" denotes an amide group whose carbonyl bond points away from the cavity of the wheel.

Supramolecular architectures of the rotaxane type have been in the focus of chemical research during the last years, because they can possibly be employed as the fundamental units of molecular machines [9, 10]. The basic idea behind this approach is a controlled motion of the wheel between different "docking stations" along the axle, which is referred to as "shuttling" [11]. In general, non-covalent axle-wheel interactions at the docking stations ensure the extra stability of the

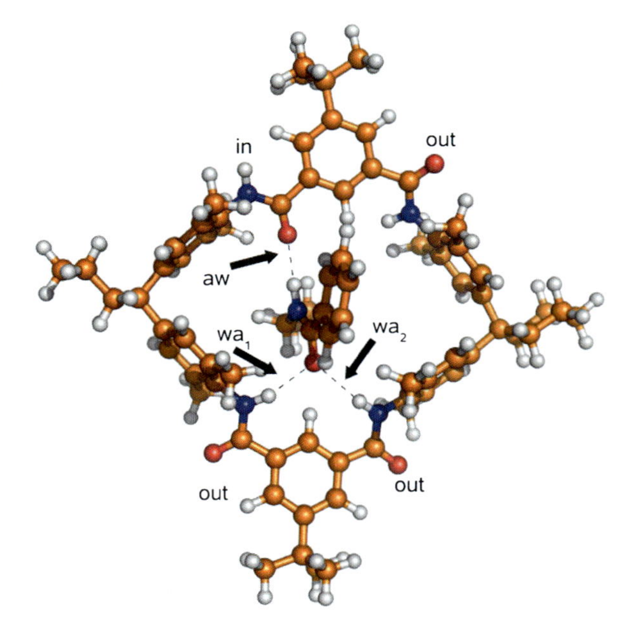

Fig. 3.1 Ball and stick model of a pseudorotaxane complex. The preorganization is realized via a template effect relying on hydrogen bonding between amide groups. Hydrogen bonds and conformations of wheel amide groups are labeled. Color code: *White* Hydrogen; *blue* Nitrogen; *red* Oxygen; *orange* Carbon

resulting complex. The controlled activation of these docking stations (e.g. via a pH gradient or redox processes) can result in a controlled shuttling process of the wheel from one docking station to another. In addition to these controlled shuttling processes, the construction of a molecular motor, i.e., a unit which converts energy into work, would be possible on the basis of an unidirectional rotation of the wheel around the axle. This principle, for instance, is employed in the enzyme atp-synthase, which transforms adp (adenosine diphosphate) and phosphate to atp (adenosine triphosphate) via a unidirectional rotation of its axle-like F_1 component [12, 13]. In this example the unidirectional rotation is ensured in terms of a proton gradient inside a channel of helical geometry, thereby introducing chirality to the system. There are several possibilities to enforce a unidirectional rotation of a macrocyclic component in an artificial supramolecular system, for instance via the exploitation of topological chirality [14]. Further details about these topics can be found in [10, 14].

3.1.2 Systems Investigated

3.1.2.1 Optimized Geometries

The test systems applied in this chapter for the assessment of the rrho model are pseudorotaxanes **V1**–**V5** consisting of the Hunter-Vögtle macrocycle **V** and axles bearing a secondary benzoylamide group but different substituents **1**–**5** at the *para* position of the aromatic ring, see Fig. 3.2 upper panel [15]. Furthermore, a sixth Vögtle-type complex (denoted as **V6**) has been considered, which does not contain one of the typical benzoylamide axles, but a single chloroform molecule. This structure will be employed for the comparison of the thermochemical reaction data obtained from the rrho approach with the corresponding experimentally determined quantities, which were measured in chloroform as solvent. In contrast to the pseudorotaxanes **V1**–**V5**, this complex only exhibits a single hydrogen bond contact between the guest CH group and the wheel carbonyl oxygen in the "in" conformation (see Fig. 3.1). In addition, some results are presented for a different kind of pseudorotaxane complexes, namely the Leigh-type pseudorotaxanes **L1′**–**L5′** composed of a characteristic disubstituted fumaramide axle **1′**–**5′** and a benzoylamide-based macrocycle **L**, see Fig. 3.2 lower panel. The computational methodology applied for the electronic structure calculations is summarized in Sect. 7.1.

As indicated in Figs. 3.1 and 3.2, the Vögtle-type pseudorotaxanes are preorganized in terms of three amide hydrogen bonds, two of which are directed from neighboring NH groups of the wheel to the axle carbonyl group (wa_1 and wa_2) and one connecting the axle NH group to a wheel carbonyl group exhibiting the "in" orientation (aw). It is also apparent from Fig. 3.2 that the axle's substituent **X** is conjugated with the carbonyl part of the amide group via the aromatic system, which is the reason why a pronounced effect of the substituent on the two

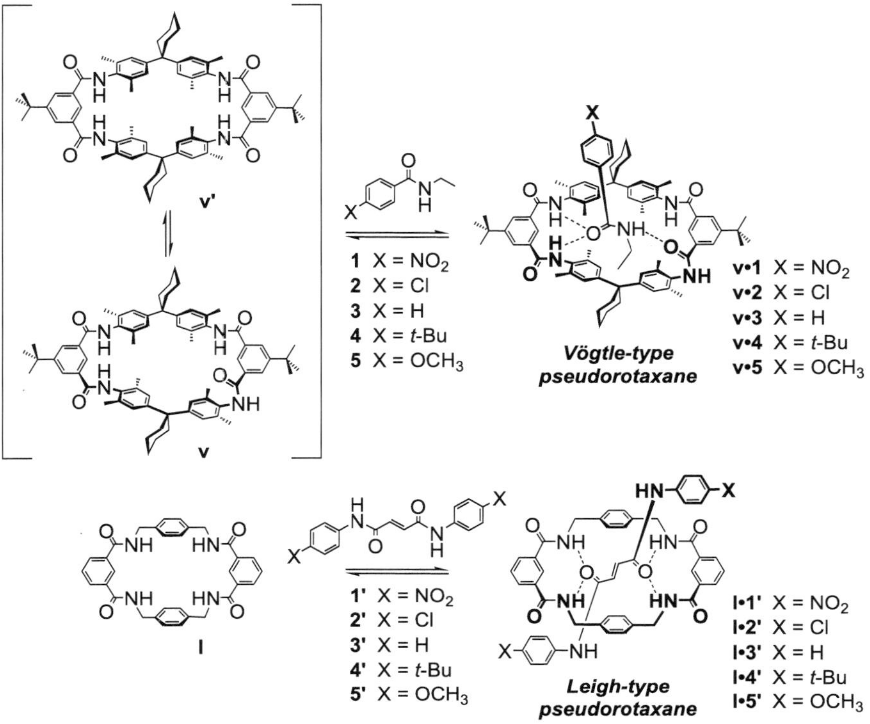

Fig. 3.2 Chemical composition of the investigated pseudorotaxane complexes and substitution pattern of the different axles. *Upper panel* Vögtle-type pseudorotaxanes; *lower panel* Leigh-type pseudorotaxanes. Reprinted with permission from Ref. [3]. Copyright 2010 American Chemical Society

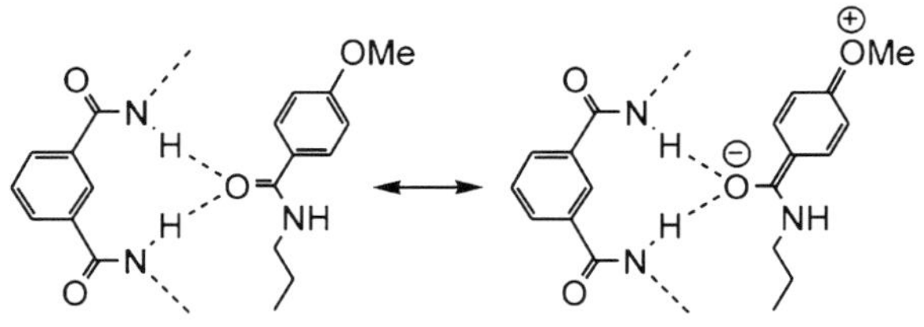

Fig. 3.3 Mesomeric effect of an electron-donating substituent on the electron density at the carbonyl oxygen atom of the axle's amide group in the Vögtle-type pseudorotaxane architecture. Spickermann et al. *How can rotaxanes be modified by varying functional groups at the axle? A combined theoretical and experimental analysis of thermochemistry and electronic effects* [1]. Copyright Wiley-VHC Verlag GmBH & Co. KGaA. Reproduced with permission

wheel-to-axle hydrogen bonds wa_1 and wa_2 could be expected. This is illustrated in Fig. 3.3 using the example of the methoxy substituent **5** [1]. Electron-donating substituents are able to increase the electron density at the axle amide group, which enables the formation of a larger charge transfer and thereby of a stronger wheel-to-axle hydrogen bond. On the contrary, if electron-withdrawing substituents are present, the negative charge at the carbonyl oxygen is expected to be smaller, which would possibly result in a weaker hydrogen bond. However, these simple empirical considerations have to be confirmed by electronic structure calculations. Due to the different atom connectivity in the axles of the Leigh-type pseudorotaxane architecture, there is no conjugation possible between the substituents and the carbonyl part of the amide group participating in the four wheel-to-axle hydrogen bonds, i.e., on the basis of this qualitative approach one would expect a rather small influence of the substituents on the strength of the hydrogen bonds in these structures.

The geometric parameters (bond lengths and angles) of the hydrogen bonds in the Vögtle-type pseudorotaxanes **V1–V5** and the chloroform complex **V6** as obtained from the geometry optimization are summarized in Table 3.1 [1]. Table 3.2 contains the same information for the Leigh-type pseudorotaxane structures [3]. Due to the C_{2h} symmetry of these complexes, all four hydrogen bonds are equivalent. Therefore, the listed parameters are averages over the four hydrogen bonds in a given complex. The numbers from Table 3.1 indicate that the bond lengths r(OH) of the different hydrogen bonds occurring in the Vögtle-type complexes differ significantly by up to 42 pm. Largest and smallest values are found for the wa_2 bond and the aw bond, respectively, and the wa_1 bond is always of intermediate length no matter which substituent is present. There are small substitution trends visible for all hydrogen bonds in the Vögtle-type pseudorotaxanes. In the case of the two wheel-to-axle hydrogen bonds wa_1 and wa_2 the bond lengths decreases for the stronger electron-donating substituents, whereas the bond length increases with decreasing electron-accepting ability of the substituent for the bond aw.

However, the observed trend is small and possibly within the error of the applied quantum chemical method. These trends are in line with the reasoning behind Fig. 3.3, from which stronger wheel-to-axle hydrogen bonds and a weaker

Table 3.1 Bond lengths r_{OH}, r_{ON} and bond angles α_{OHN} of all three hydrogen bonds for complexes **V1–V5** as well as the corresponding values for the CH····O hydrogen bond found for complex **V6** [1]

No.	Guest	wa_1			wa_2			aw		
		r_{OH}	r_{ON}	α	r_{OH}	r_{ON}	α	r_{OH}	r_{ON}	α
V1	R-NO_2	218	316	163.2	244	344	159.6	202	301	162.0
V2	R-Cl	215	314	163.6	242	341	160.7	204	303	162.2
V3	R-H	215	314	164.1	241	340	161.2	206	304	159.4
V4	R-^{t}Bu	215	314	164.2	239	339	161.6	207	304	159.8
V5	R-OCH_3	213	312	162.0	239	338	164.5	208	306	160.5
V6	$CHCl_3$	–	–	–	–	–	–	202	311	172.2

All distances in [pm] and all angles in [°]. Abbreviations: *w* wheel; *a* axle (guest)

Table 3.2 Hydrogen bond lengths r_{OH}, r_{ON} and angles α_{OHN} of the Leigh-type complexes **L1′**–**L5′** (C_{2h} symmetry) [3]

No.	Guest	r_{OH}	r_{ON}	α
L1′	R-NO_2	229	328	162.8
L2′	R-Cl	224	323	165.4
L3′	R-H	222	322	166.1
L4′	R-^{t}Bu	221	322	166.9
L5′	R-OCH_3	220	321	167.9

The listed values are averages over all four hydrogen bonds in every structure. All distances in [pm] and all angles in [°]

axle-to-wheel hydrogen bond are expected in the case of electron-donating substituents.

With regard to this point it should be noted that a correlation between the length and the strength of a bond in general is not necessarily given, but in the present case apparently does exist, see also the following part [16]. Identical trends are found for the r(ON) bond lengths, which indicates a nearly undisturbed NH bond length of approximately 99–100 pm (the NH bond length in the isolated wheel and axles is equal to 100 pm). There is no clear trend present in the calculated hydrogen bond angles, which exhibit values of approximately 160°. The hydrogen bond length in the Vögtle-type solvent complex **V6** is comparable to those for the aw hydrogen bond in the corresponding pseudorotaxanes, but the angle α(OHC) is larger by approximately 10°.

The calculated geometries for the Leigh-type pseudorotaxanes show a trend of the hydrogen bond length which is similar to the one observed for the wa_1 hydrogen bond in the Vögtle-type complexes, but slightly more pronounced. This is remarkable, since one might expect a stronger influence of the substituent in the Vögtle-type structures due to its conjugated arrangement to the carbonyl group. This observation indicates that in the present case the disubstitution of the Leigh-type axles has a more significant effect on the hydrogen bond situtation as the conjugated connectivity in the case of the Vögtle-type axles. The values obtained for the oxygen–nitrogen distances again point at a nearly undisturbed NH bond length, and the angles are also similar to the values of the Vögtle-type structures, but show a reversed trend with regard to the corresponding bond lengths as compared to the situation found for the wa_1 bond, see Table 3.1. This behavior can again be rationalized if the bond length is taken as an indicator for the bond strength, in which case the stronger bonds of the axles substituted with electron-donating groups tend to assume a hydrogen bond geometry close to the optimal arrangement of 180°.

3.1.2.2 Interaction Energies

In addition to obtaining the optimized structures discussed in the previous paragraphs, the results of the electronic structure calculation can be evaluated in analogy to Eq. 2.50, thereby yielding the supramolecular interaction energies ΔE_{tot}

Table 3.3 Different energies for the Vögtle-type complexes

No.	Guest	Individual hydrogen bond energy			Total interaction energy	
		$E^{sen}_{wa_1}$	$E^{sen}_{wa_2}$	E^{sen}_{aw}	ΔE^{tot}	ΔE^{tot}_{ZPE}
V1	R-NO_2	−13.0	−3.8	−16.0	−35.7	−32.6
V2	R-Cl	−14.4	−4.4	−14.8	−36.6	−33.9
V3	R-H	−14.4	−4.6	−13.9	−36.5	−33.9
V4	R-^{t}Bu	−14.4	−5.0	−13.4	−36.0	−34.6
V5	R-OCH_3	−15.8	−5.2	−13.1	−37.9	−35.8
V6	$CHCl_3$	–	–	–	−6.8	−6.5

Individual hydrogen bond energies for wa_1, wa_2, and aw are given in addition to the total counterpoise corrected interaction energies ΔE^{tot} including the zero point energy correction ΔE^{tot}_{ZPE} [1]. All energies in [kJ/mol]

which are a direct measure for the total interaction between wheel and axle.[2] Table 3.3 lists these energy differences for the Vögtle-type pseudorotaxanes together with the energies of the individual hydrogen bonds obtained from the shared electron number (SEN) analysis (see Sect. 7.1 for the computational methodologies) [17]. It should be noted that the values in Table 3.3 are approximately corrected for the basis set superposition error in terms of the counterpoise correction approach [18, 19]. The overall interaction between wheel and axle as indicated by the ΔE^{tot} values is of moderate strength and lies between −35 and −38 kJ/mol for all substituents. The interaction energy of the solvent complex **V6** is significantly smaller (−6.8 kJ/mol) due to the presence of only a single hydrogen bond. There is only a small trend visible in these interaction energies, which indicates more stable complexes for the electron-donating substituents. This can be rationalized as in the case of the hydrogen bond lengths discussed before, i.e., these substituents have a stabilizing effect on the two wheel-to-axle bonds wa_1 and wa_2 as well as a destabilizing effect only on the single axle-to-wheel bond aw, compare also Table 3.1. The total interaction energy corrected for the zero point energy differs by approximately 2–3 kJ/mol from the values without this correction as apparent from the values in column ΔE^{tot}_{ZPE} in Table 3.3, thereby indicating an insignificant contribution from this nuclear quantum effect on the overall interaction in the investigated complexes.

A more detailed examination of the energetic circumstances in the hydrogen bonds between wheel and axle can be achieved in terms of the individual hydrogen bond energies $E^{sen}_{wa_1}$, $E^{sen}_{wa_2}$, and E^{sen}_{aw} as obtained from the shared electron number analysis.[3] These numbers show clear trends with respect to the substitution pattern, which parallel the tendencies observed for the bond lengths of the individual

[2] As far as all interactions relevant in the system under study are adequately described by the applied quantum chemical method.

[3] Please note that the sum of the individual interaction energies $E^{sen}_{wa_1}$, $E^{sen}_{wa_2}$, and E^{sen}_{aw} is always of smaller magnitude than the total interaction energy ΔE^{tot}, since the latter one includes additional host–guest interactions as e.g. dipole–dipole interactions in contrast to the SEN analysis which solely accounts for the actual bond under consideration.

hydrogen bonds in Table 3.1. The two wheel-to-axle hydrogen bonds wa_1 and wa_2 are of significantly different strength and differ by approximately 10 kJ/mol, no matter which substituent is present. The strength of the third hydrogen bond aw is comparable to the wa_1 bond, but follows an inverse trend as indicated in the discussion of Fig. 3.3. Both wheel-to-axle hydrogen bonds are stabilized by electron-donating substituents to a similar extent (the wa_2 bond is stabilized to a lesser extent), thereby justifying the picture of an increased electron density at the amide group of the axle as discussed previously. In addition, this explanation is also consistent with the reversed trend observed for the axle-to-wheel hydrogen bond aw. This bond is stabilized by electron-withdrawing substituents which are able to reduce the electron density at the amide group, thereby increasing the positive charge at the hydrogen bond donor atom. Such a charge increase results in a pronounced charge transfer between hydrogen bond acceptor and donor and finally in a stronger hydrogen bond. The observed trends for the wheel-to-axle and axle-to-wheel hydrogen bonds also explain the relatively small trend in the interaction energies ΔE^{tot}, since these values capture the total interaction between wheel and axle and therefore include both compensating effects. Thus, the negligible substituent effect on the overall host–guest interaction in the Vögtle-type pseudorotaxanes is rooted in the different nature of the hydrogen bonds between the two components.

These considerations are also supported by a calculation of atomic charges on the basis of the natural population analysis (NPA) summarized in Table 3.4 [1, 20]. In the case of the two wheel-to-axle hydrogen bonds wa_1 and wa_2, the negative charge on the axle carbonyl oxygen atom is increased by electron-donating substituents as indicated in Fig. 3.3, thereby stabilizing these two interactions. In contrast, the positive charge on the axle amide hydrogen atom involved in the axle-to-wheel bond is increased by electron-withdrawing substituents, which leads to a stronger axle-to-wheel interaction in agreement with the results presented in Table 3.3. However, the substituent effect is considerably smaller in the latter case, which could be based on the larger distance between the substituent and the hydrogen bond donor of the axle-to-wheel interaction, see Fig. 3.2.

The individual as well as the total interaction energies for the Leigh-type pseudorotaxanes are summarized in Table 3.5. The total interaction energies ΔE^{tot}

Table 3.4 Results of the natural population analysis for complexes **V1–V5** [1]

No.	Guest	wa_1		wa_2		aw	
		q^{don}	q^{acc}	q^{don}	q^{acc}	q^{don}	q^{acc}
V1	R-NO_2	0.450	−0.747	0.445	−0.747	0.463	−0.712
V2	R-Cl	0.452	−0.755	0.447	−0.755	0.460	−0.707
V3	R-H	0.453	−0.758	0.448	−0.758	0.459	−0.703
V4	R-^{t}Bu	0.453	−0.761	0.449	−0.761	0.458	−0.701
V5	R-OCH_3	0.454	−0.765	0.449	−0.765	0.457	−0.703

Values for the charge analysis refer to the hydrogen atom (q^{don}) and hydrogen bond acceptor (q^{acc}). For denotation of hydrogen bonds see Fig. 3.1. All charges are in [e]

Table 3.5 Individual hydrogen bond energies E^{sen} and total interaction energies ΔE^{tot} including zero point energy correction ΔE^{tot}_{ZPE} for the Leigh-type pseudorotaxanes [3]

No.	Guest	E^{sen}	ΔE^{tot}	ΔE^{tot}_{ZPE}
L1′	R-NO_2	−6.8	−35.7	−29.5
L2′	R-Cl	−8.8	−46.4	−39.0
L3′	R-H	−9.3	−51.4	−43.9
L4′	R-^{t}Bu	−9.6	−54.2	−46.7
L5′	R-OCH_3	−10.7	−56.4	−47.8

Please note that due to the symmetry all four hydrogen bonds are equivalent. Therefore only the average value for E^{sen} is given. All energies in [kJ/mol]

exhibit larger absolute values, but are still in the same order of magnitude as the numbers obtained for the Vögtle-type structures. In contrast to the Vögtle-type complexes the Leigh-type pseudorotaxanes contain disubstituted axles in which no conjugation is present between the substituents and the amide group forming the hydrogen bond to the wheel, see Fig. 3.2. In addition, all hydrogen bonds in the Leigh-type structures are of the same kind, i.e., all four hydrogen bonds are wheel-to-axle interactions in which the NH part of the four wheel amide groups acts as the hydrogen bond donor and the carbonyl part of the axle amide groups as hydrogen bond acceptor.

This different bonding situation is clearly reflected in the numbers listed in Table 3.5. The total interaction energies ΔE^{tot} show a well-defined trend which is not present in the corresponding energies of the Vögtle-type pseudorotaxanes (see Table 3.3), even though no direct conjugation between substituent and the hydrogen bond location is possible in the Leigh-type structures, see Fig. 3.2. As in case of the wheel-to-axle hydrogen bonds in the Vögtle-type structures (see third and fourth column in Table 3.3), electron-donating substituents stabilize the total interaction between wheel and axle, whereas electron-withdrawing substituents have a destabilizing effect. This observation is again in agreement with the picture of a higher electron density at the axle carbonyl oxygen atoms in case of electron-donating substituents and thereby with a stabilization of the corresponding hydrogen bond. Due to the missing axle-to-wheel hydrogen bond, no compensating effects as in the case of the Vögtle-type structures occur. Consequently, this trend is also present in the individual hydrogen bond energies obtained in terms of the SEN analysis, see the third column in Table 3.5. As in the case of the individual energies for the wheel-to-axle hydrogen bonds in the Vögtle-type complexes, a stabilization of each individual hydrogen bond by the presence of electron-donating substituents is observed, but for a given substituent the two strong hydrogen bonds in the Vögtle-type structures (wa_1 and aw) are always more stable than one of the hydrogen bonds in the Leigh-type complexes. However, in the case of the Leigh-type structures the net stabilization obtained by changing the substituent from NO_2 to OCH_3 is larger for the total interaction ($\sim$20 kJ/mol) and for the individual hydrogen bond ($\sim$4 kJ/mol) as compared to the corresponding stabilization in the Vögtle-type pseudorotaxanes ($\sim$2 and $\sim$3 kJ/mol, respectively). The reason for this relatively large stabilization of the Leigh-type complexes besides the absence of counterbalancing trends in the individual hydrogen bonds is thus possibly found in the disubstitution of the fumaramide axles. Furthermore, it is interesting to see that the gradual stabilization of the total host–guest interaction ΔE^{tot} assumes largest values for the replacement of NO_2 by Cl ($\sim$10 kJ/mol) and Cl by H ($\sim$5 kJ/mol), i.e., for the reduction of the electron-withdrawing character of the substituent. The increase of the electron-donating character of the substituent (i.e., changing **L3′** to **L4′** and **L4′** to **L5′**) results in a smaller stabilization of $\sim$3 and $\sim$2 kJ/mol, respectively. The more pronounced substituent effect in the Leigh-type pseudorotaxanes is also visible in the influence of the zero point energy correction to the total interaction energy (see the last column in Table 3.5), which approximately amounts up to 10 kJ/mol for the methoxy substituent. Thus, clear trends are apparent in the interaction energies predicted for the Leigh-type pseudorotaxanes, even though no direct conjugation according to Fig. 3.3 is possible in these structures.

3.1.3 Evaluation of Thermodynamic Quantities

3.1.3.1 Models for the Pseudorotaxane Formation

According to the explanation presented in Sect. 2.1.2, the computation of the rigid rotor harmonic oscillator partition function Q can be carried out on the basis of electronic structure calculations, i.e., if the structural (mass and principal moments of inertia) and vibrational (harmonic frequencies) properties of the system under investigation are known. The calculation of thermodynamic quantities from the partition function Q is routinely achieved following textbook procedures, see for instance Eqs. 2.60 and 2.64 [21]. However, the values obtained in this way are in most cases unsuitable for a direct comparison with experimental measurements, since they represent absolute values with a reference temperature of $T = 0$ K. The experimental free energy changes employed in this chapter for the evaluation of the rrho approach were obtained via a ^{1}H NMR titration analysis of the association reaction between the free wheel and the free axle in chloroform solution [1]. In the frame of the rrho approach this association between wheel and axle can be modelled in analogy to the computation of the supramolecular (or total) interaction energy as expressed in Eq. 2.50, i.e., the value of a given quantity (e.g. the free energy G) calculated for the isolated wheel and the isolated axle are subtracted from the corresponding value calculated for the pseudorotaxane complex. The change in the quantity under examination obtained from this course of action would then represent the association reaction according to the direct formation [1]

$$\text{axle} + \text{wheel} \longrightarrow \text{pseudorotaxane}. \tag{3.1}$$

A more elaborate approach would consider the fact that the experimental reference values were obtained in chloroform solution via some form of microsolvation, which explicitly includes a limited number of solvent molecules in the electronic structure calculation. Since the association reaction takes place in a localized area near the cavity of the wheel, negligible effects from solvent molecules at the outside of the macrocycle can be expected. However, it is reasonable to expect one (or more) solvent molecules to occupy the cavity itself, because the formation of a hydrogen bond between the carbonyl groups of the wheel and the chloroform molecule is energetically favorable, see the last line of Table 3.3. In this case the association reaction between wheel and axle could be understood as a displacement process in the cavity of the wheel, in which the chloroform guest is replaced by the axle. In order to model this guest exchange reaction in terms of the rrho approach, electronic structure calculations for the complex **V6** containing a single chloroform molecule inside the cavity of the wheel were carried out. The changes in the different thermodynamic quantities are then obtained by subtracting the corresponding values for the wheel including the solvent molecule and the axle from the sum of the values obtained for the appropriate pseudorotaxane complex and the released chloroform molecule. Formally, this corresponds to the following reaction equation [1]

$$\text{axle} + \text{wheel} \cdot CHCl_3 \longrightarrow \text{pseudorotaxane} + CHCl_3. \qquad (3.2)$$

In Eq. 3.2 the term "wheel $\cdot$ $CHCl_3$" denotes the solvent complex **V6**. The two different approaches summarized in Eqs. 3.1 and 3.2 will be referred to as "direct formation" and "exchange formation", respectively, and they will both be applied for the calculation of thermodynamic reaction data in the following parts. In the case of the exchange reaction defined by Eq. 3.2, the possibility of a second chloroform guest in the cavity of the solvent complex **V6** at reaction start was considered additionally. However, the changes in the quantities obtained from using this extended complex as the reactant instead of structure **V6** are relatively small and do not lead to an improvement with respect to the experimentally determined values [1]. Therefore, the **V6** structure will exclusively be employed for the calculation of thermodynamic quantities according to the exchange formation reaction in Eq. 3.2.

3.1.3.2 Thermodynamic Quantities for the Direct Formation

In this part the focus is set to the pseudorotaxane association according to the direct formation (see Eq. 3.1) as introduced in the last paragraph. Table 3.6 summarizes the corresponding thermodynamic energy changes calculated for

Table 3.6 Thermochemical quantities at $T = 298.15$ K and $p = 101{,}325$ Pa according to the direct formation of Eq. 3.1 [1]

No.	Guest	ΔH	$T\Delta S$	ΔG
V1	R-NO_2	−26.9	−52.1	25.1
V2	R-Cl	−28.1	−50.6	22.4
V3	R-H	−28.2	−50.4	22.2
V4	R-^{t}Bu	−28.2	−49.5	21.3
V5	R-OCH_3	−29.7	−49.0	19.3

All values in [kJ/mol]

standard conditions ($T = 298.15$ K and $p = 101{,}325$ Pa) in terms of the rrho approximation. The change in enthalpy ΔH amounts -25 to -30 kJ/mol and is negative for all investigated substituents, thereby indicating a favorable process from the pure enthalpic point of view. This observation is to be expected, since the formation of hydrogen bonds between wheel and axle upon association yields an energetic stabilization (see also Table 3.3).

It is also obvious that the small substitution trend observed in the thermally uncorrected total interaction energies ΔE^{tot} in Table 3.3 is present in the enthalpy changes as well. Thus, the considerations pointed out in the preceding paragraphs regarding the increase of electron density at the atomic sites relevant for the host–guest interaction by electron-donating substituents are still valid at standard conditions, at least according to the rrho model. As in the case of the total interaction energies, the largest relative substituent effects on ΔH are found for the strongest electron-withdrawing and the strongest electron-donating substituent. However, the values calculated for ΔH also indicate that the overall interaction between wheel and axle is weakened by approximately 10 kJ/mol as compared to the zero Kelvin interaction energies ΔE^{tot} listed in Table 3.3.

The calculated association entropies $T\Delta S$ reveal a significant loss of entropy at $T = 298.15$ K upon formation of the pseudorotaxane complex. There is a substitution trend present in the entropy changes as well, but this trend inverts the one found for the interaction energies and enthalpies. The magnitude of the entropy loss is almost twice as large as the change in the enthalpy for the electron-withdrawing substituents and is still considerably larger as the enthalpy contribution for the electron-donating substituents, thereby clearly indicating a disfavoring entropic effect upon host–guest association. Taken together, these results finally lead to positive values for the free energy change of association in the range of 19–25 kJ/mol, i.e., the formation of all investigated Vögtle-type pseudorotaxanes is predicted to be unstable in thermodynamic terms by the direct formation model. Due to the reversed substitution trend found for the enthalpy and entropy contributions as well as the different sign of these two quantities in the Gibbs–Helmholtz equation, a larger relative stability is calculated for the complexes bearing electron-donating substituents as already observed in the case of the total interaction energies, see Table 3.3. However, each individual complex is predicted to be unstable at standard conditions according to the direct formation model.

A more detailed analysis of the calculated thermodynamic association quantities can be carried out by considering the effect of the different molecular degrees of freedom. In the frame of the rrho approach, the single particle partition function q is factorized into contributions from the molecular degrees of freedom, and these contributions are available as analytic functions of molecular properties, see Sect. 2.1.2. These analytic expressions enable the decomposition of most thermodynamic quantities into additive contributions from the different degrees of freedom, which represents an advantage of the simple rrho approach over many of the other approaches for the calculation of thermodynamic quantities presented in Chap. 2 [22]. The contributions from the translational, rotational, and vibrational degrees of freedom for the axle-wheel association to the Vögtle-type complexes according to the direct formation model are listed in Table 3.7. It should be noted that there is no explicit contribution from the electronic degrees of freedom, since all quantum chemical calculations were carried out for the electronic ground state and the zero of energy is set to the energy of the isolated wheel and axle at an infinite separation, i.e., only the first term in Eq. 2.43 is considered. This choice for the zero of energy also implies that the contribution to the change in enthalpy and free energy due to the change in electronic structure upon complex formation is directly given by the adiabatic interaction energies ΔE^{tot} (see Table 3.3), i.e., the enthalpy change ΔH in Table 3.6 corresponds to the total interaction energy ΔE^{tot} according to Eq. 2.50 and the contributions from the changes due to the remaining degrees of freedom (translation, rotation, vibration). These latter values are the ones which are listed in Table 3.7.

From these numbers it is apparent that the translational and rotational degrees of freedom yield system-unspecific enthalpy contributions of −3.7 kJ/mol. This value corresponds to the one expected from the equipartition theorem of classical thermodynamics, which states that in thermal equilibrium each degree of freedom entering the Hamiltonian quadratically has an average energy of $(3/2)RT$ [22]. At standard conditions, this is equal to 3.7 kJ/mol, and according to Eq. 3.1 the change in enthalpy is thus given as −3.7 kJ/mol. The fact that the classical result is obtained even though the rrho approach is a model relying on quantum statistics can be understood by considering the different approximations this approach relies on, see Sect. 2.1.2. For the analytical evaluation of the translational and rotational partition functions, the summations in Eqs. 2.27 and 2.34 are replaced by integrals

Table 3.7 Thermochemical quantities at $T = 298.15$ K and $p = 101{,}325$ Pa according to the direct formation of Eq. 3.1 [1]

No.	Guest	ΔH	$T\Delta S$
Translation			
V1	R-NO_2	−3.7	−51.6
V2	R-Cl	−3.7	−51.4
V3	R-H	−3.7	−50.8
V4	R-^{t}Bu	−3.7	−51.7
V5	R-OCH_3	−3.7	−51.3
Rotation			
V1	R-NO_2	−3.7	−40.4
V2	R-Cl	−3.7	−39.9
V3	R-H	−3.7	−38.7
V4	R-^{t}Bu	−3.7	−41.0
V5	R-OCH_3	−3.7	−39.9
Vibration			
V1	R-NO_2	16.2	39.9
V2	R-Cl	15.9	40.7
V3	R-H	15.8	39.1
V4	R-^{t}Bu	15.2	43.2
V5	R-OCH_3	15.6	42.2

All values in [kJ/mol]

over the corresponding quantum numbers. Due to this high temperature approximation, the discrete quantum mechanical character of translation and rotation is lost and these degrees of freedom are treated in a pure classical sense, i.e., their discrete energy spectra are replaced by an energy continuum. Consequently, the obtained translational and rotational enthalpy contributions are equal to the values obtained from classical thermodynamics.[4] This is different in the case of the vibrational enthalpy contributions, since the spacing between vibrational energy levels in general is too large to be approximated by a continuum, see the discussion of Eq. 2.40. Therefore, these contributions considerably depend on the number and the frequency of the involved vibrational degrees of freedom. However, there is only a small variation visible in the vibrational enthalpy changes listed in the last block of Table 3.7, which can be attributed to the fact that upon complex association only a relatively small number of modes is affected directly e.g. due to hydrogen bond formation. The positive sign of the vibrational contributions can be explained by the fact that there is a larger number of vibrations present in the associated pseudorotaxanes, i.e., the vibrational enthalpy of the product is larger than that of the reactants. Compared to the translational and rotational contributions, the numbers obtained for the vibrations are relatively large. This is the reason why the combined enthalpy contribution from translation, rotation, and vibration is also positive and thereby yields a destabilizing contribution to the pseudorotaxane association. However, this unfavorable effect is more than compensated by the formation of the three hydrogen bonds (and additional host–guest interactions) as expressed in the energies from Table 3.3. There is also a slight substituent trend apparent in the vibrational enthalpy changes, which indicates the differing effect of electron-withdrawing and electron-donating substituents on the electron density and thereby on the vibrational modes involving the hydrogen bonding sites. In combination with the frequency changes occurring upon complex formation, such effects can be exploited for a correlation between substituent effect, frequency shift, and hydrogen bond energy as demonstrated in recent studies [2, 3].

The situation for the entropy contributions arising from the translational, rotational, and vibrational degrees of freedom is comparable to the behavior found for the corresponding enthalpy contributions only concerning the influence of the substituent as well as the sign. The values predicted for the translational association entropy contribution are not substituent-specific and lie in a narrow interval between −50 and −52 kJ/mol at the chosen conditions. Compared to the translational enthalpy change, these values assume a considerably larger magnitude. This behavior can be attributed to the different number of translational degrees of freedom on both sides of the reaction arrow in the direct formation model, see Eq. 3.1. The same is true in the case of the rotational entropy contributions, which parallel the corresponding translational numbers but are smaller in absolute value

[4] It should be noted that this reasoning is only valid for the enthalpy contributions, since there is no equipartition theorem for the entropy.

by approximately 11 kJ/mol. The unspecific character for these two types of degrees of freedom with regard to the substituent can be ascribed to the fact that the effect of the substituent on the relevant molecular quantities (mass and principal moments of inertia, see Eqs. 2.30 and 2.37) is present in the reactants as well as in the associated product, thereby affecting the entropy change for the association only in an insignificant manner. As in the case of the enthalpy, the vibrational entropy changes show a different sign as compared to the translational and rotational entropy contributions due to the larger number of vibrations in the pseudorotaxane complexes. For each investigated complex, the magnitude of the vibrational entropy change matches the one for the corresponding rotational contribution to within 2–3 kJ/mol, which results in a virtual cancellation of the entropy changes arising from these two degrees of freedom. The total entropy change upon complex association therefore is almost completely determined by the translational contribution, which explains the similarity of the entropy changes in Table 3.6 and the first block of Table 3.7. In addition, the magnitude of the translational entropy changes is always considerably larger than those of the rotational and vibrational contributions at least in the case of all investigated pseudorotaxane complexes. The generality of these observations will be examined in Sect. 3.2.

The thermodynamic situation obtained from the rrho approach and the direct formation model thus predicts an increase of the free energy upon the association of the wheel and all axles and according to that an association constant smaller than one [23]. This thermodynamic instability of the pseudorotaxane complexes is not affected by the substitution pattern, but instead arises due to a large entropy loss upon complex association.

3.1.3.3 Thermodynamic Quantities for the Exchange Formation

The exchange formation reaction as expressed in Eq. 3.2 models the pseudorotaxane association in terms of a guest displacement process taking place in the macrocycle's cavity. Upon the formation of the pseudorotaxane complex, a solvent molecule previously bound to the macrocycle is released. The thermodynamic quantities calculated for this process according to the rrho approach are summarized in Table 3.8. The computed enthalpy changes ΔH are again negative and very similar to the values obtained for the direct formation reaction in Table 3.6. In addition, the small substituent trend found in the case of the total interaction energies (see Table 3.3) is again reflected in the enthalpy changes.[5] Thus, the transition from the direct formation to the exchange formation has virtually no

[5] It should be noted that the contribution to the association enthalpy due to the change in total energy in the case of the exchange formation is given as the ΔE^{tot} value for the pseudorotaxane complex *minus* the ΔE^{tot} value for the solvent complex **V6**. In the case of the direct formation this contribution directly corresponds to the ΔE^{tot} value of the pseudorotaxane complex. The difference is based on the distinct form of the reaction equations, see Eqs. 3.1 and 3.2.

Table 3.8 Thermochemical quantities at $T = 298.15$ K and $p = 101{,}325$ Pa according to the exchange formation of Eq. 3.2 [1]

No.	Guest	ΔH	$T\Delta S$	ΔG
V1	R-NO_2	−27.1	−10.1	−17.0
V2	R-Cl	−28.3	−8.6	−19.7
V3	R-H	−28.4	−8.5	−19.9
V4	R-^{t}Bu	−28.4	−7.5	−20.9
V5	R-OCH_3	−29.9	−7.1	−22.8

Complex **V6** ($CHCl_3$) is used as reaction partner for modeling the exchange reaction. All values in [kJ/mol]

effect on the total association enthalpies. This observation is not necessarily to be expected, since the change in total energy upon complex association according to the exchange formation differs from the one for the direct formation by the negative interaction energy of the solvent complex **V6** (+6.8 kJ/mol, see Table 3.3). However, this difference is almost completely compensated by a difference in association enthalpy of reversed sign (−7.0 kJ/mol in case of complex **V5**), which results in a negligible net effect in the enthalpy change between the two different association models.

The situation is completely reversed in the case of the association entropies $T\Delta S$. The entropy changes computed for the exchange formation reaction still indicate a loss of entropy during the exchange process and thereby a destabilizing contribution to the change in free energy, but the magnitude of this entropy loss is significantly smaller compared to the enthalpy change of the exchange formation as well as the entropy change of the direct formation, see Table 3.6. As in the case of the association entropies of the direct formation, there is a small substituent trend present in the $T\Delta S$ values listed in Table 3.8, which covers an interval of approximately 3 kJ/mol and again is reversed to the trend found for the corresponding enthalpy changes. However, the striking difference lies in the much smaller absolute value of the predicted entropy loss upon complex formation. Compared to the computed enthalpy changes, the entropy change is smaller by approximately 17 kJ/mol in the case of the nitro-substituted complex **V1** and by more than 20 kJ/mol in the case of the methoxy-substituted compound **V5**. These considerable differences are also present in the quantities obtained for the direct formation, but in that case the magnitude of the entropy change is *larger* by the corresponding amount.

Consequently, this reduced entropy contribution is also visible in the calculated free energy changes. All investigated Vögtle-type pseudorotaxane complexes are predicted to be stable in thermodynamic terms by the exchange formation model as indicated by the negative sign of the computed free energy changes. There is no difference concerning the substitution trend between the two models, though. Both the direct formation and the exchange formation predict the association of axle **5** and wheel **V** to the methoxy-substituted complex **V5** to be more stable by $\Delta\Delta G = -5.8$ kJ/mol as compared to the nitro-substituted compound **V1**. This trend is also consistent with the one found for the zero Kelvin interaction energies ΔE^{tot} and for the individual hydrogen bond energies $E^{sen}_{wa_1}$, $E^{sen}_{wa_2}$, and E^{sen}_{aw} listed in

Table 3.9 Thermochemical quantities at $T = 298.15$ K and $p = 101{,}325$ Pa according to the exchange formation of Eq. 3.2 [1]

No.	Guest	ΔH	$T\Delta S$
Translation			
V1	R-NO_2	0.0	−1.8
V2	R-Cl	0.0	−1.6
V3	R-H	0.0	−1.0
V4	R-^{t}Bu	0.0	−1.9
V5	R-OCH_3	0.0	−1.5
Rotation			
V1	R-NO_2	0.0	−6.3
V2	R-Cl	0.0	−5.8
V3	R-H	0.0	−4.6
V4	R-^{t}Bu	0.0	−6.9
V5	R-OCH_3	0.0	−5.8
Vibration			
V1	R-NO_2	1.3	−2.0
V2	R-Cl	1.0	−1.2
V3	R-H	0.9	−2.8
V4	R-^{t}Bu	0.3	1.3
V5	R-OCH_3	0.7	0.3

Complex **V6** ($CHCl_3$) is used as reaction partner for modeling the exchange reaction. All values in [kJ/mol]

Table 3.3. These analogies indicate that both thermodynamic models support the simple picture of a stabilizing effect due to increased charge transfer via the two wheel-to-axle hydrogen bonds wa_1 and wa_2 in the presence of electron-donating substituents as indicated in Fig. 3.3. However, a final assessment of the two proposed approaches can only be achieved through a comparison to the real thermodynamic situation obtained from experimental measurements. This validation will be presented in the next section.

The contributions of the translational, rotational, and vibrational degrees of freedom to the association enthalpy and entropy according to the exchange formation model are summarized in Table 3.9. In the case of the enthalpy the listed values are again the net thermal corrections without the change in total energy as for the corresponding values in Table 3.7, see the discussion of these values in the last part. In contrast to the direct formation, there is no change in the translational and rotational enthalpies upon complex association. This is to be expected, since these numbers are computed in terms of the rrho approximation, and in this setup the translational and rotational degrees of freedom are treated in a classical way as discussed for the translational and rotational enthalpy changes of the direct formation.

Thus, each particle taking part in the reaction yields a constant contribution of $(3/2)RT$, and due to the constant particle number on both sides of the reaction arrow (see Eq. 3.2) the net enthalpy change is zero. This is different in the case of the vibrational enthalpy contributions, which again assume positive values and show a small, irregular substituent dependancy as found for the corresponding numbers of the direct formation in Table 3.7. The difference in vibrational association enthalpy computed for the nitro-substituted complex **V1** and the methoxy-substituted complex **V5** amounts only 0.6 kJ/mol, which exactly corresponds to the value found for this difference in the case of the direct formation, and complex **V4** is identified as an outlier with respect to this trend. The situation predicted for the translational, rotational, and vibrational contributions to the association entropy is different. The translational contributions are negative and thereby disfavor the complex association as in the case of the direct formation. However, due to the small and substituent-unspecific magnitude this destabilizing effect is only of minor importance in contrast to the considerable destabilization found in the case of the direct formation, see the first block in Table 3.7. The rotational association entropies are still significantly smaller than the corresponding numbers of the direct formation reaction, but they exceed the translational as well as the vibrational contributions and constitute the most important part of the overall entropy change upon complex association. This is in contrast to the situation observed for the direct formation, where the translational degrees of freedom constitute the most important contribution and the rotational and vibrational parts cancelled each other almost completely. A possible explanation for this increased importance could lie in the enlarged relative effect of the substituent on the molecular moments of inertia in the isolated axles as compared to the associated pseudorotaxane and resulting from that a larger rotational partition function, compare Eq. 2.37. This assumption is supported by the fact that complex **V4** bearing the substituent with the largest mass also shows the largest loss of rotational entropy upon association, and that the unsubstituted compound **V3** yields the smallest rotational entropy change due to the small mass of the hydrogen atom. Clearly, this effect will also be present in the direct formation model, but in that case the influence of the changing number of translational degrees of freedom seems to be even larger. There is no substituent trend visible in either the translational or rotational association entropies in accordance with the absence of such a trend in the direct formation model. The calculated vibrational association entropies are again of a smaller magnitude and do not show a clear substituent trend as well. However, there is a change in the sign of these contributions for the electron-donating substituents in **V4** and **V5**. The corresponding entropy changes calculated for the direct formation (see Table 3.7) are positive and of considerable magnitude, which can clearly be attributed to the larger number of vibrational degrees of freedom in the associated complex. In the present case, there is no change in the number of normal modes upon complex association and more subtle effects due to the different masses or steric demands of the substituents become important. This is supported by the fact that the largest loss of vibrational entropy is predicted for the smallest "substituent" in **V3** and that the largest gain of vibrational entropy is obtained for the largest and most heavy substituent in **V4**. However, the exact nature of these influences is difficult to estimate from electronic structure calculations.

At last, the thermodynamic picture emerging from the exchange formation model is in strong contrast to the situation obtained from the direct formation model. Whereas there is virtually no difference in the association enthalpies predicted from both models, the difference in the predicted association entropies is always larger than 40 kJ/mol at standard conditions for all investigated systems. Even though the same substituent trends are present in both models, the absolute differences are considerably larger and in the case of the exchange formation lead to an association process which is favorable in thermodynamic terms due to a relatively small entropy loss upon complex formation. Thus, the pseudorotaxane formation reaction is predicted to be an exergonic process by the exchange formation and to be an endergonic process by the direct formation.

3.1.3.4 Comparison to the Experiment and Model Assessment

The results presented in the previous parts demonstrate that the two proposed models for the pseudorotaxane association reaction make a contrary prediction concerning the thermodynamic stability of the investigated pseudorotaxanes. Thus, the experimental examination of the real thermodynamic situation accompanying the association reaction would clearly help to discriminate between the actual physical significance of the two models. The last column in Table 3.10 summarizes the experimental free energies of association measured in ^{1}H NMR titration experiments at $T = 303$ K and standard pressure [1]. From these numbers it is apparent that the association reaction between wheel and axle in chloroform solution is an exergonic process at the chosen temperature and pressure conditions as predicted by the exchange formation model. In addition, the small stability trend with regard to the substituent predicted by both models as well as the zero Kelvin interaction energies in Table 3.3 is not present in the experimental values. Given that the error bar of the applied experimental method amounts to ±2 kJ/mol, a discussion of the very small substitution effects found in the ΔG_{exp} values would not be reasonable [1, 24].

In fact, the important information to be extracted from the ΔG_{exp} values lies in the observation that all investigated axles form a stable complex with the macrocycle **V** and that the magnitude of the free energy lowering for this process is

Table 3.10 Calculated thermochemical quantities at $T = 298.15$ K and $p = 101{,}325$ Pa as well as the experimental free energy change of association ΔG_{exp} measured at $T = 303$ K [1]

No.	Guest	Direct formation			Exchange formation			Experiment
		ΔH	$T\Delta S$	ΔG	ΔH	$T\Delta S$	ΔG	ΔG_{exp}
V1	R-NO_2	−26.9	−52.1	25.1	−27.1	−10.1	−17.0	−13.7
V2	R-Cl	−28.1	−50.6	22.4	−28.3	−8.6	−19.7	−13.6
V3	R-H	−28.2	−50.4	22.2	−28.4	−8.5	−19.0	−11.0
V4	R-^{t}Bu	−28.2	−49.5	21.3	−28.4	−7.5	−20.9	−11.4
V5	R-OCH_3	−29.7	−49.0	19.3	−29.9	−7.1	−22.8	−12.1

Complex **V6** ($CHCl_3$) is used as reaction partner for all other guests in case of the exchange formation. All values in [kJ/mol]

considerably larger than the errors of the applied methodology, thereby allowing a robust determination of the thermodynamic stability of all investigated pseudo-rotaxanes. This result also demonstrates in an unambiguous way that the exchange formation is the only one of the two proposed models being able to predict the qualitatively correct thermodynamic situation of the pseudorotaxane association in solution. The comparison of the free energy changes calculated from this model to the experimental ones shows that the calculated values overestimate the stability of all pseudorotaxanes by approximately 5–10 kJ/mol. The smallest differences are found for the complexes bearing the electron-withdrawing substituents, which indicates that either the amount of electron donation of the corresponding substituents in **V4** and **V5** is overestimated by the applied electronic structure method or that these effects are weakened by e.g. solvent effects more subtle than the single chloroform molecule used for the microsolvation in the formation exchange model. However, considering the trends found in the calculated enthalpy and entropy contributions in the second block of Table 3.10, it is seen that the larger discrepancies in free association energy calculated for the electron-donating substituents arise from *both* the enthalpic as well as the entropic part. In addition to a more negative enthalpy change in the case of the complexes bearing electron-donating substituents, a smaller loss of entropy upon complex formation is predicted for these compounds, thereby contributing to their increased stability in terms of the free energy change. Additional experimental measurements have been carried out according to the van't Hoff method for the determination of enthalpy and entropy changes in the case of complex **V2** [24]. The results of these experiments yield an enthalpy contribution of $\Delta H_{\text{exp}} = -22.0$ kJ/mol and an entropy contribution of $T\Delta S = -8.8$ kJ/mol (at $T = 303$ K) to the association reaction [1]. Upon comparison with the numbers in the second block of Table 3.10, these numbers indicate that the association entropy is very well captured by the exchange formation model and that the discrepancies in the free association energy are due to an overestimation of the association enthalpy magnitude alone. These observations thereby indicate that it is possible to model the entropy change of reactions in the condensed phase according to the rrho approach quite accurately if a reasonable model of the process under investigation is taken as a basis, which is an important result, since condensed phase entropies and entropy changes are often considered to be more difficult to calculate from theoretical approaches than e.g. enthalpy changes [25]. On the other hand, the deviation in the association enthalpy could well be based on the computed total interaction energies from Table 3.3 and thereby on the quality of the applied quantum chemical methodology, since in addition to neglecting certain molecular interactions an overestimation of interaction energies by dft methods has been reported in the literature before [26, 27]. Thus, the prediction of the free energy change of complex association in solution could be improved even further by applying a more sophisticated method for the electronic structure calculation, as for example Møller–Plesset perturbation theory. However, due to the large number of atoms in the investigated pseudorotaxane complexes such a method refinement would also involve a significant increase in the computational effort, but in the case of smaller systems such an approach would seem reasonable. A more elaborate electronic structure method could even be able to capture the electronic situation in the complexes bearing electron-donating substituents (**V4** and **V5**) more accurately and thereby reduce the larger discrepancies found for these compounds in the computed association enthalpies, see Table 3.10.

The comparison to the experimental free association energies clearly demonstrates that only the exchange formation is able to predict the correct thermodynamic behavior of the pseudorotaxane association, and the discussion laid out in the two previous parts as well as the summary in Table 3.10 clearly indicate that the thermodynamic difference between the two proposed models is based on entropic contributions alone. From this point of view the change in entropy upon complex association in solution is described much more accurately by a displacement of a solvent molecule from the wheel cavity as by an exclusive binding of the axle to an empty macrocycle. This result could have been expected from chemical knowledge due to the high concentration of the solvent present in the solution. However, the relevant methodological difference between the direct formation and the exchange formation is certainly given by the number of particles on the sides of the reactants and the products, respectively. This is immediately seen from Eqs. 3.1 and 3.2. The particle number on the side of the reactants is always two, but in the case of the direct formation the two reactants form a single product, whereas the release of the chloroform molecule in the case of the exchange formation leads to the occurrence of a second particle on the product side of the reaction equation. Besides the low probability for an unoccupied cavity at liquid phase densities, the reaction course according to Eq. 3.1 seems to be artificial in a methodological sense as well. The reason for that can be found in the entropy contributions from the different molecular degrees of freedom summarized in Table 3.7. Compared to the corresponding contributions calculated for the exchange formation model (see Table 3.9), the entropy changes for each individual type of degree of freedom are considerably larger in the case of the direct formation model. In addition, the entropy change is controlled by the translational part alone due to the almost complete compensation of the vibrational and rotational parts. Consequently, these large individual contributions must therefore be based on the change in particle number occurring in the direct formation reaction. This conclusion is rationalized in a straightforward way by counting the single degrees of freedom of each type on both sides of the reaction equation. In the case of the direct formation, six translational and rotational degrees of freedom in the reactants are compared to only three translational and rotational degrees of freedom in the product. Thus, the "loss" of these degrees of freedom explains the loss of entropy predicted for the translational and rotational contributions by the direct formation. Of course, these degrees of freedom are not lost but are merely transformed to additional vibrational modes in the product, which explains the gain in entropy upon complex association predicted for the vibrational degrees of freedom. However, the gain in entropy due to the newly established normal modes cannot account for the significant combined loss in entropy due to the translational and rotational degrees of freedom as indicated by the numbers in Table 3.7, which finally results in a large unfavorable entropy change for the association reaction. Such an interconversion of different types of molecular degrees of freedom does not occur in the exchange formation model. Due to the uniform number of particles on both sides of the reaction equation, the number of translational, rotational, and vibrational degrees of freedom remains constant as well, and the comparison to the experimentally determined values demonstrates that this situation is the physically more reasonable variant in the high density regime of the liquid phase. However, the question remains why the loss of entropy arising from translation and rotation cannot be compensated by the vibrational degrees of freedom, in which case the total entropy change would be closer to the real thermodynamic situation found for the complex association. In order to clarify this matter, a closer look on the artificial decomposition into these degrees of freedom and the accuracy of the resulting entropy contributions at liquid phase densities has to be taken, which will inevitably lead back to the factorization of the molecular partition function in the rrho model as expressed in Eq. 2.26. A detailed analysis of these aspects will be presented in the next section. The results of the actual section have shown that the differing influence of the translational, rotational, and vibrational degrees of freedom on the entropy can be bypassed by keeping the number of particles constant during the course of the reaction. Thus, the formulation of model reactions according to that rule should yield reasonable entropy changes as found for the systems investigated in this chapter and can be recommended for modeling entropy changes in the condensed phase according to the rrho approach.

3.2 A Quantitative Error Analysis of the Rigid Rotor Harmonic Oscillator Model

3.2.1 Analysis of the Particle Number Effect

The results and conclusions presented in the previous sections demonstrate that the stoichiometry of the reaction for which thermochemical reaction data according to the rrho protocol are calculated has a considerable effect on the entropy contributions. If the number of particles does not stay constant during the reaction, an interconversion of translational and rotational degrees of freedom into vibrational modes (or vice versa) takes place, which in the general case of an association reaction between compound A and compound B to the complex A···B results in the formation of six new vibrations and the loss of three translations as well as three rotations according to

$$\underbrace{A+B}_{\text{6trans/rot}} \longrightarrow \underbrace{A\cdots B}_{\text{3trans/rot}} . \tag{3.3}$$

The decomposition of the entropy changes for the complex association according to the direct formation (see Table 3.7) has shown that the translational part has the largest influence, while the rotational and vibrational parts cancel each other almost exactly. In order to confirm these observations and to exclude coincidental effects due to the specific binding situation in the pseudorotaxane structures the results of additional calculations for the association of small molecules will be presented in the next paragraph. The comparison of the calculated values to experimental association entropies measured in the gas phase will show if the pronounced particle number effect arises from the application of the simple rrho model to systems at liquid phase densities or if it is an artificial contribution due to approximations (or errors) in the applied methodology. From the results presented so far it is clear that the change in particle number during the reaction course only affects the entropy changes and not the change in enthalpy (see Table 3.10). Therefore, the focus will be set on this quantity for the rest of the chapter.

3.2.1.1 Model Reactions in the Gas Phase

The examination of the agreement between thermochemical quantities of gas phase reactions and the corresponding predictions made by the rrho approach will provide information about the performance of the model at the conditions it was originally developed for, namely the low density gas phase. It is reasonable to expect that the various approximations introduced in the derivation of the rrho approach (see Sect. 2.1.2) are more accurate at high temperatures and low densities, but these conditions are still different from the isolated molecule picture the rrho approach is based on. In order to quantify these differences in thermodynamic terms, a comparison to experimental measurements is essential. Furthermore, the comparison of association entropies measured in the gas phase to the association entropy of complex **V2** ($T\Delta S = -8.8$ kJ/mol at $T = 303$ K) in solution will provide an order-of-magnitude estimate for the effect of the increased density on the entropy change upon complex formation in real systems. The selected gas phase association reactions are the dimerization of water,

$$2\mathrm{H_2O} \longrightarrow (\mathrm{H_2O})_2, \tag{3.4}$$

the dimerization of acetic acid,

$$2\mathrm{H_3CCOOH} \longrightarrow (\mathrm{H_3CCOOH})_2, \tag{3.5}$$

and the dimerization of nitrogen dioxide,

$$2\mathrm{NO_2} \longrightarrow \mathrm{N_2O_4}. \tag{3.6}$$

All of these reactions are association reactions and therefore are subject to the particle number effect in the rrho model. However, the type of association is different in each case. The dimerization of the two water molecules is accompanied by the formation of a single hydrogen bond, whereas two hydrogen bonds are established during the association of the acetic acid monomers. Thus, it is reasonable to expect that the flexibility of the dimer with regard to the isolated monomers is smaller in the case of acetic acid, which is the reason why a larger loss of entropy should be expected for that system. The same is true for the association of nitrogen dioxide, upon which a covalent bond is formed in contrast to the non-covalent interactions in the two former cases.

The association entropies as obtained from the rrho approach at $T = 298.15$ K as well as the experimentally determined values are summarized in Table 3.11 [31]. From these numbers it is apparent that for all investigated systems the change in translational entropy lies between −40 kJ/mol and approximately −45 kJ/mol at standard conditions. Given the large difference in mass, system size, and general complexity of the system, these numbers are very similar to the translational association entropy predicted for the direct formation of the pseudorotaxane complexes, see Table 3.7. Larger discrepancies between the examined reactions are found in the case of the rotational and vibrational association entropies. The loss of rotational entropy upon formation of the water dimer ($T\Delta S = -3.3$ kJ/mol) is considerably smaller than the corresponding value calculated for the nitrogen dioxide dimerization ($T\Delta S = -16.4$ kJ/mol), although both complexes have a comparable spatial extent. However, this observation can be rationalized due to the smaller mass of the hydrogen atoms and the reduced influence on the principal moments of inertia in the complex as compared to the terminal oxygen atoms in the hydrazine molecule. The largest magnitude of rotational association entropy is predicted for the dimerization of the large acetic acid molecule, and the comparison of the rotational contributions in Table 3.11 to the ones from Table 3.7 indicates that the change in rotational entropy is significantly affected by the size and the mass of the investigated system, respectively. The situation is again different in the case of the vibrational association entropies. As in the case of the pseudorotaxane complexes, these contributions are positive due to the formation of six new vibrational modes upon complex association, but are considerably smaller as compared to the numbers summarized in Table 3.7. In addition, the vibrational entropy changes from Table 3.11 show differences of up to 8 kJ/mol, which are not present in the corresponding numbers calculated for the pseudorotaxanes. These large differences relative to the total magnitude of the vibrational

Table 3.11 Contributions from translational, rotational, and vibrational degrees of freedom to the association entropy for three gas-phase reactions according to the rrho approach at $T = 298.15$ K and $p = 101{,}325$ Pa

Reaction	$T\Delta S_{trans}$	$T\Delta S_{rot}$	$T\Delta S_{vib}$	$T\Delta S_{el}$	$T\Delta S_{tot}$	$T\Delta S_{tot}^{exp}$
$2H_2O \rightarrow (H_2O)_2$	−40.6	−3.3	12.9	0.0	−31.0	−23.2 ± 1.7
$2AcOH \rightarrow (AcOH)_2$	−45.1	−23.6	20.7	0.0	−48.0	−44.3 ± 1.3
$2NO_2 \rightarrow N_2O_4$	−44.1	−16.4	12.1	−3.4	−51.8	−52.7

Experimental data and corresponding error bars are given for comparison (no error bars are provided in [28]) [28–30]. All values are in [kJ/mol]

contributions apparently do not correlate with changes in the mass, the strength of interaction, or the loss of flexibility upon complex formation as could be rationalized in the case of the rotational contributions. Consequently, the compensation between the rotational and vibrational association entropies observed for all pseudorotaxane complexes (see Table 3.7) is not present in the case of the smaller complexes in Table 3.11. However, the largest effect on the net association entropy still comes from the translational degrees of freedom. A novel feature introduced by the nitrogen dioxide dimerization is the contribution of the electronic degree of freedom to the association entropy. The reason for this contribution is directly visible from Eqs. 2.43 to 2.64 and lies in the fact that the nitrogen dioxide monomer is an open shell system with a single unpaired electron. The multiplicity of the electronic ground state is therefore given as $g_1 = 2$, and according to that the entropy of the electronic ground state is equal to $S_{el} = R\ln(2)$, which results in an entropy loss of -3.4 kJ/mol at $T = 298.15$ K for the dimerization due to the formation of a closed-shell system in the hydrazine molecule. The calculated net association entropies $T\Delta S_{tot}$ predict a disfavoring entropic contribution for all three gas-phase dimerizations amounting between -31 and -52 kJ/mol, and the largest contribution to this negative association entropy in all cases stems from the translational part, which constitutes -40 to -45 kJ/mol and thereby overrules the contributions from rotations and vibrations. This behavior is not based on the rather compact structure of the chosen molecules, since it is also observed for the much larger and more widespread pseudorotaxane complexes, see Table 3.7. As expected, the vibrational contribution to the association entropy partially compensates for the unfavorable contributions, but in all three cases it cannot account for the rotational part alone though the difference between $T\Delta S_{vib}$ and $T\Delta S_{rot}$ for the acetic acid and nitrogen dioxide dimerization is rather small. Due to the large vibrational contribution and the small magnitude of the rotational association entropy computed for the water dimerization the magnitude of the net entropy loss is rather small as compared to reactions 3.5 and 3.6, which show comparable values. The numbers computed for the latter two reactions (-48.0 and -51.8 kJ/mol, respectively) are again very similar to the ones predicted for the pseudorotaxane association, see Table 3.6. Due to the considerable differences in structure as well as bonding situation between these systems, this finding indicates a universality of rrho calculated association entropies of a rather system-unspecific nature. The dimerization of water in the gas phase constitutes an obvious exception to that observation.

Of course it is no surprise that an association in the gas phase is accompanied by a formal loss of entropy, which is in accordance with the picture of a higher ordered reaction product. The question to be addressed here is whether this loss of entropy is properly reproduced by the methods of the rrho approach. The experimental dimerization entropies at $T = 298.15$ K for the investigated reactions are equal to $T\Delta S^{exp}_{H_2O} = -23.2 \pm 1.7$kJ/mol, $T\Delta S^{exp}_{NO_2} = -52.7$kJ/mol (no error bars given), and $T\Delta S^{exp}_{AcOH} = -44.3 \pm 1.3$ kJ/mol [28–30]. Upon comparison with the theoretically predicted values it is quite surprising that the putatively most simple system, namely the water dimer, exhibits the largest deviation between the prediction according to the rrho model and the experiment, and that the computed nitrogen dioxide dimerization entropy including the open-shell spin contribution probably is as accurate as the experimental error bars. The small deviations found for the acetic acid and nitrogen dioxide association entropies again demonstrate that size and molecular complexity do not necessarily give an indication for the accuracy of the resulting rrho-computed entropy changes, and more important that the degree-of-freedom interconversion inherent in the dimerization reactions is accurately described in terms of the rrho model in *gas phase* reactions.

With the exception of the water dimerization reaction, the association entropies computed for reactions of the form as in Eq. 3.3 according to the rrho approach lie in a rather narrow interval around −50 kJ/mol at $T = 298.15$ K, see Tables 3.6 and 3.11. The experimental values listed in Table 3.11 indicate that an entropy loss of that magnitude is a realistic prediction for association reactions in the gas phase. In contrast, the experimentally determined association entropy for the pseudorotaxane formation of compound **V2** amounts to $T\Delta S = -8.8\,\text{kJ/mol}$ ($T = 303$ K) [1]. From these numbers, a difference in association entropy of approximately $T\Delta\Delta S = 40\,\text{kJ/mol}$ between gas phase association and association in solution can be estimated. This is of course only an order-of-magnitude estimate, since the entropy change for pseudorotaxane **V2** is the only reference value for an association in solution and the universality of this value could be rather low due to specific solvent-solute interactions or the complexity of the involved structures. However, a model for the prediction of association entropies in the condensed phase on the basis of the rrho approach has to account for a difference of this order of magnitude upon transition from the low density to the high density domain. The considerations from Sect. 2.1.2 show that the density dependancy in the rrho approach is established by applying the model of the particle in a box to the translational degrees of freedom, thereby introducing the volume of free translation (the box volume) to the formalism. This is in agreement with the observations of the present chapter that the translational degrees of freedom have the largest effect on the entropy changes of association reactions according to Eq. 3.3.

3.2.1.2 Entropy Contributions Due to the Interconversion of Different Degrees of Freedom

In order to quantify the observed difference between the translational, rotational, and vibrational degrees of freedom on the association entropies in the gas phase as well as in the liquid phase, the formal relation between the mechanical quantity relevant for each degree of freedom (volume, mass, principal moments of inertia, frequencies) and the corresponding entropy contribution are required. In complete analogy to the course of action from Sect. 2.2.2, these are obtained by exploiting the factorization of the rrho partition function according to Eq. 2.26 as well as the logarithmic relation between entropy and partition function as expressed in

Eq. 2.64. Due to this logarithmic dependancy, the total entropy according to Eq. 2.64 can be decomposed into additive contributions from the different degrees of freedom according to

$$
\begin{aligned}
S &= k_B \ln\left(\frac{q^N}{N!}\right) + k_B T\left(\frac{\partial \ln(q^N/N!)}{\partial T}\right)_{N,V} \\
&= k_B \left[\ln\left(q_{trans}^N q_{rot}^N q_{vib}^N q_{el}^N\right) - \ln(N!)\right] + k_B T\left(\frac{\partial \ln(q_{trans}^N q_{rot}^N q_{vib}^N q_{el}^N)}{\partial T}\right)_{N,V} \\
&= \left[Nk_B \ln(q_{trans}) + Nk_B T\left(\frac{\partial \ln(q_{trans})}{\partial T}\right)\right] + \left[Nk_B \ln(q_{rot}) + Nk_B T\left(\frac{\partial \ln(q_{rot})}{\partial T}\right)\right] \\
&\quad + \left[Nk_B \ln(q_{vib}) + Nk_B T\left(\frac{\partial \ln(q_{vib})}{\partial T}\right)\right] + \left[Nk_B \ln(q_{el}) + Nk_B T\left(\frac{\partial \ln(q_{el})}{\partial T}\right)\right] \\
&\quad - k_B \ln(N!),
\end{aligned}
\tag{3.7}
$$

where q_{trans}, q_{rot}, q_{vib}, and q_{el} denote the rrho partition functions of the corresponding degrees of freedom as introduced in Sect. 2.1.2. However, care has to be taken of the $N!^{-1}$ factor arising in the (N, V, T) partition function due to the indistinguishability of the particles (see Eq. 2.26), which only occurs once and therefore has to be assigned to one of the degree-of-freedom dependent partition functions. This $-k_B \ln(N!)$ term in the entropy decomposition is usually resolved according to Stirling's approximation and combined with the translational partition function q_{trans} (see Eq. 2.30), which results in the following expression for the translational entropy

$$
\begin{aligned}
S_{trans} &= Nk_B \ln(q_{trans}) + Nk_B T\left(\frac{\partial \ln(q_{trans})}{\partial T}\right)_{N,V} - Nk_B \ln(N) + Nk_B \\
&\overset{N=N_A}{\Longrightarrow} R \ln\left(\frac{V}{N_A}\left[\frac{2\pi m k_B T}{h^2}\right]^{3/2}\right) + \frac{5}{2}R.
\end{aligned}
\tag{3.8}
$$

This equation is the well-known Sackur–Tetrode equation for the molar entropy of a classical ideal gas, which is to be expected, since the translational degrees of freedom are the only degrees of freedom in such a system [21]. The evaluation of the rotational and vibrational contributions in Eq. 3.7 according to Eqs. 2.37 and 2.41 leads to the following molar entropy contributions from these degrees of freedom [21]

$$
S_{rot} = R \ln\left(\frac{\pi^{1/2}}{\sigma}\left[\frac{T^3}{\Theta_A \Theta_B \Theta_C}\right]^{1/2}\right) + \frac{3}{2}R \tag{3.9}
$$

$$
S_{vib} = R \sum_{k=1}^{3M-6} \frac{\beta h \nu_k}{\exp(\beta h \nu_k) - 1} - \ln[1 - \exp(-\beta h \nu_k)], \tag{3.10}
$$

where all symbols have the same meaning as in Sect. 2.1.2. An interesting detail in these individual contributions is the sum occurring in the vibrational entropy in Eq. 3.10, which indicates that each single mode of vibration yields an independent contribution to the entropy solely depending on the temperature and the frequency of that mode. This is of course a direct consequence of the transformation to normal coordinates and the decoupling of the vibrational modes resulting from that, which leads to the product in the polyatomic harmonic oscillator partition function (see Eq. 2.41) and due to the logarithmic dependancy between entropy and partition function to the sum in Eq. 3.10. In contrast, the translational and rotational entropies inherently include the three degrees of freedom of the respective kind, as can be seen from the product of the three rotational temperatures in Eq. 3.9 and the occurrence of the volume in Eq. 3.8. Based on this observation, a further decomposition of the vibrational contributions listed in Table 3.11 for the gas phase dimerization of the small molecules is possible. Assuming that the dimer modes which are already present in the monomers are not shifted by large amounts upon dimerization, the vibrational association entropies will be a direct consequence of the six new vibrational modes formed in the complexes due to the loss of three translational and three rotational degrees of freedom according to Eq. 3.3. The validity of this assumption can be checked by the numbers from Table 3.12, which summarize the effect of the new vibrational modes in terms of wavenumbers and entropy contributions. The first block in Table 3.12 gives the wavenumbers of the vibrational modes which are formed in the dimers during the association process due to the loss of three translational and three rotational degrees of freedom.

However, this process will also have an effect on the modes already present in the monomers, and the magnitude of this effect can be estimated by the mean square shift $\Delta\nu_{\text{shift}}$ of the remaining vibrations. This value is calculated according to

Table 3.12 Wave numbers ν and vibrational entropies TS of the six new vibrations in the dimers

Frequencies							
Dimer	ν_1	ν_2	ν_3	ν_4	ν_5	ν_6	$\Delta\nu_{\text{shift}}$
$(H_2O)_2$	164	175	202	216	404	645	29
$(AcOH)_2$	63	89	108	168	194	201	27
N_2O_4	88	190	262	400	452	636	33
Vibrational entropies							
Dimer	TS_1	TS_2	TS_3	TS_4	TS_5	TS_6	$T\Delta S_{\text{shift}}$
$(H_2O)_2$	3.1	3.0	2.6	2.5	1.2	0.5	0.0
$(AcOH)_2$	5.4	4.6	4.1	3.1	2.7	2.7	−1.9
N_2O_4	4.6	2.8	2.1	1.2	1.0	0.5	0.2

$\Delta\nu_{\text{shift}}$ denotes the mean square shift of the remaining vibrational modes due to the dimerization weighted by the total number of these modes. $T\Delta S_{\text{shift}}$ gives the combined entropy contribution of these vibrational shifts. All entropies are calculated at $T = 298.15$ K and $p = 101{,}325$ Pa. All wavenumbers in [cm^{-1}] and all entropies in [kJ/mol]

$$\Delta \nu_{\text{shift}} = \frac{1}{N_{\text{vib}}} \sqrt{\sum_{k}^{N_{\text{vib}}} \Delta \nu_k^2}, \tag{3.11}$$

where $\Delta \nu_k$ denotes the shift in wavenumber for each individual mode k and N_{vib} the number of the vibrational degrees of freedom in the dimers *without* the six modes introduced through the association process. The numbers listed in the first block of Table 3.12 indicate that most of the six new vibrations in the dimers occur in the low wavenumber domain below 450 cm^{-1}. In fact, with the exception of the acetic acid dimer the newly formed modes exhibit wavenumbers considerably lower than the lowest modes of the corresponding monomers, and the difference between the lower monomer modes and the new vibrations is almost always larger than several hundreds of wave numbers in the examined dimers. In the case of acetic acid, there are two low-lying modes in the monomer as well, which correspond to bending motions of the methyl group. The examination of the normal coordinates belonging to the new vibrational degrees of freedom in the dimers demonstrates that these modes are mainly related to bending vibrations of the whole complex which can be understood as hindered translations and rotations, thereby indicating their origin from these degrees of freedom. The smallest wavenumbers below 100 cm^{-1} are obtained for the acetic acid dimer and the hydrazine molecule, whereas the lowest vibration in the water dimer lies at $\nu = 164\ \text{cm}^{-1}$. This rather large difference can be rationalized in terms of the small mass of the water molecule as compared to the one of acetic acid and nitrogen dioxide. The values calculated for the mean square shift of the remaining frequencies are in comparable ranges of approximately 30 cm^{-1} per mode, which indicates a small, but non-negligible effect of the dimerization on the modes of the monomers on average. The second block in Table 3.12 summarizes the effects on the entropy due to the changed vibrational situation in the dimers. According to Eq. 3.10, low-frequency vibrations yield contributions of largest magnitude to the vibrational entropy (see also the plot of S_{vib} in Sect. 3.2.2). This important observation is clearly visible in the entropies in Table 3.12. The largest individual entropy per mode is found for hydrazine as well as the dimer of the acetic acid, which also show the lowest individual mode ν_1 in terms of wavenumbers. However, the wavenumbers of the subsequent vibrations $\nu_2 - \nu_6$ rise considerably faster in the case of hydrazine as compared to the dimers of water and acetic acid, and considerable contributions to the vibrational entropy arise from the first three modes alone. This is different in the case of the water dimer and especially in the case of the acetic acid dimer. In the latter case, all newly formed modes assume wavenumbers below approximately 200 cm^{-1}. Hence, all of these new vibrations contribute in a significant manner to the total vibrational entropy, which is clearly reflected in the large vibrational entropy gain of the acetic acid dimer upon association as listed in Table 3.11. In the case of the water dimer, the relatively low wavenumbers of the first four modes result in significant vibrational entropy contributions, and in combination with the larger wavenumber of the first mode the overall situation of the vibrational entropy is comparable to the one of the hydrazine dimerization, see Table 3.11.

In addition to the new vibrational modes established during the dimerization process, the shift of the modes already present in the monomers contribute to the vibrational association entropy as well. This effect is quantified by the $T\Delta S_{\text{shift}}$ values in Table 3.12, which represent the difference in entropy of a given mode in the dimer and the same mode in the monomer summed over all modes already present in the corresponding monomers. It is important to note that these entropy contributions cannot be obtained from the frequency shifts of the corresponding modes in a reasonable way, because the vibrational entropy is very sensitive to the magnitude of the frequency (as can be seen from the individual mode entropies in Table 3.12) and more important because S_{vib} is not defined for negative shifts (i.e., if the mode is shifted to a lower wavenumber in the dimer) due to the logarithmic term in Eq. 3.10. By calculating the absolute entropy and taking the difference between the value of a certain mode in the dimer and in the monomer instead, shifts of existing modes which lead to lower wavenumbers result in positive entropy contributions and vice versa, thereby providing a reliable measure of the frequency shift effect on the vibrational entropy arising from modes already present in the monomers.

In the case of the three investigated model reactions this contribution is qualitatively different for each case, but the magnitude is always negligible as compared to the effect of the new vibrational modes formed during the dimerization. With the exception of very large shifts (which are not observed in these systems), the magnitude of the shifts arising from the dimerization of water is always insignificant due to the large wavenumbers of the three modes in the water monomer (ranging from approximately 1,600–3,800 cm^{-1}). As long as the shifted mode is in that order-of-magnitude wavenumber domain, the effective vibrational entropy contribution is much smaller than the accuracy of the computations and negligible compared to the other factors contributing to the entropy change, which results in a zero net effect from the shifts of existing vibrational modes in the case of water. A similar situation is observed for the dimerization of nitrogen dioxide, for which the only entropy contribution worthy of mention arises due to the shift of the lowest mode in the nitrogen dioxide monomer (742 cm^{-1}) by approximately -15 cm^{-1}, thereby leading to an insignificant $T\Delta S_{\text{shift}}$ value of 0.2 kJ/mol. The corresponding value calculated for the dimerization of acetic acid equals -1.9 kJ/mol and is thus larger in magnitude as the previously discussed numbers, but still smaller in absolute value than the entropy contribution from every single of the newly formed modes. This increase in magnitude as compared to the $T\Delta S_{\text{shift}}$ values of hydrazine and the water dimer can be rationalized in terms of the larger overall number of vibrations in the acetic acid molecule as well as the shift of the two low-lying vibrations already present in the monomer. As a consequence of the dimerization reaction, one of these vibrations is shifted from 148 to 167 cm^{-1}. In combination with a larger number of shifts to higher wavenumbers, a formal loss of entropy due to the shifts in already existing modes is observed, but as stated previously these contributions are negligible compared to the vibrational entropy gain from the newly established modes.

The association entropy contributions arising from the rotational and translational degrees of freedom are always negative due to the loss of three degrees of freedom in each case as can be seen from the corresponding numbers in Table 3.11. In contrast to the vibrational entropy change, the entropy contributions from these degrees of freedom cannot be uniformly decomposed into each single degree of freedom of the respective kind. It is therefore more instructive to obtain equations for the rotational and translational entropy change for reaction equations of the form as in Eq. 3.3 by evaluating the corresponding equation

$$\Delta S_{\mathrm{rot/trans}} = S^{AB}_{\mathrm{rot/trans}} - S^{A}_{\mathrm{rot/trans}} - S^{B}_{\mathrm{rot/trans}} \tag{3.12}$$

on the basis of the absolute entropies as expressed in Eqs. 3.9 and 3.8. If the symmetry number σ of the dimer AB is the same as those of the monomers A and B, this procedure yields for the rotational association entropy

$$\begin{aligned} \Delta S_{\mathrm{rot}} &= -R \ln\left(\frac{\pi^{1/2}}{\sigma}\left[\frac{T^3}{\Delta\Theta^{A,B}_{AB}}\right]^{1/2}\right) - \frac{3}{2}R \\ &= \Delta S^{\dagger}_{\mathrm{rot}} - \frac{3}{2}R, \end{aligned} \tag{3.13}$$

where $\Delta\Theta^{A,B}_{AB}$ denotes the fraction between the rotational temperatures of the monomers A, B and the rotational temperature of the dimer AB according to

$$\Delta\Theta^{A,B}_{AB} = \frac{\Theta^A_a\Theta^A_b\Theta^A_c \times \Theta^B_a\Theta^B_b\Theta^B_c}{\Theta^{AB}_a\Theta^{AB}_b\Theta^{AB}_c}. \tag{3.14}$$

By applying the same course of action to the change in entropy due to the translational degrees of freedom the corresponding equation for $\Delta S_{\mathrm{trans}}$ is given by

$$\begin{aligned} \Delta S_{\mathrm{trans}} &= -R \ln\left(\frac{V}{N_A}\left[\frac{2\pi\Delta m^{A,B}_{AB} k_{\mathrm{B}} T}{h^2}\right]^{3/2}\right) - \frac{5}{2}R \\ &= -R \ln\left(\frac{V}{N_A}\Lambda(\Delta m)^{-3}\right) - \frac{5}{2}R, \end{aligned} \tag{3.15}$$

where $\Delta m^{A,B}_{AB}$ abbreviates the mass relation between the dimer and the monomer masses in complete analogy to the fraction of the rotational temperatures in Eq. 3.14

$$\Delta m^{A,B}_{AB} = \frac{m^A m^B}{m^{AB}}, \tag{3.16}$$

and $\Lambda(\Delta m)$ the thermal de Broglie wavelength according to Eq. 2.31 with regard to the mass relation Δm expressed in Eq. 3.16. It is apparent from Eqs. 3.13 and 3.15

that the change in rotational and translational entropy upon the dimerization consists of a constant contribution ($-(3/2)R$ and $-(5/2)R$, respectively) as well as a contribution due to the change in the relevant mechanical quantity (principal moments of inertia and mass, respectively). The multiplicative relations in Eqs. 3.14 and 3.16 found for these latter contributions are a direct consequence of the logarithmic dependancy of the entropy on these quantities, see Eqs. 3.9 and 3.8. In addition, the inverse of the number density $\rho_n^{-1} = (N_A/V)^{-1}$ yields a contribution to the translational entropy change as well. With regard to a rrho-based model for association entropies in the high density domain, a separation of the mass contribution and the density contribution to the translational entropy change would be reasonable. However, a further partitioning of the translational entropy change in Eq. 3.15 according to these considerations would result in an argument of the logarithm function not being free of units, which is not reasonable from a physical point of view. In order to solve that problem and to obtain a sensible partitioning, the introduction of an inverse unit number density (i.e., a unit volume) $\rho_n^{\dagger-1} = 1\mathrm{m}^3$ and a unit thermal de Broglie wavelength $\Lambda_1 = 1\mathrm{m}$ is necessary. These quantities can then be employed for the partitioning of the translational association entropy into a mass contribution ΔS_{mass} and a density contribution ΔS_{dens} according to

$$\begin{aligned}
\Delta S_{\mathrm{mass}} &= -R\ln\left(\rho_n^{\dagger-1}\Lambda(\Delta m)^{-3}\right)\\
\Delta S_{\mathrm{dens}} &= -R\ln\left(\frac{V}{N_A}\Lambda_1^{-3}\right)\\
\Delta S_{\mathrm{trans}} &= \Delta S_{\mathrm{mass}} + \Delta S_{\mathrm{dens}} - \frac{5}{2}R.
\end{aligned} \tag{3.17}$$

The different contributions to the rotational and translational association entropy according to Eqs. 3.13 and 3.17 are summarized in Table 3.13. From these numbers it is seen that the change in rotational entropy $T\Delta S^{\dagger}_{\mathrm{rot}}$ due to the difference in rotational temperature between dimer and the two monomers $\Delta\Theta^{A,B}_{AB}$ is rather small and positive in the case of the water dimerization and that the overall rotational association entropy is controlled by the constant contribution $(3/2)R$. This result is based on the relatively large change in the rotational

Table 3.13 Mass ($T\Delta S_{\mathrm{mass}}$) and number density ($T\Delta S_{\mathrm{dens}}$) contributions to the translational association entropy as well as rotational association entropy ($T\Delta S^{\dagger}_{\mathrm{rot}}$) without the constant contribution $(3/2)R$ (listed separately) of gas phase dimerizations at $T = 298.15$ K and $p = 101{,}325$ Pa

Reaction	$T\Delta S^{\dagger}_{\mathrm{rot}}$	$-T(\frac{3}{2}R)$	$T\Delta S_{\mathrm{mass}}$	$T\Delta S_{\mathrm{dens}}$	$-T(\frac{5}{2}R)$
$2H_2O \rightarrow (H_2O)_2$	0.4	−3.7	−179.3	144.9	−6.2
$2AcOH \rightarrow (AcOH)_2$	−19.9	−3.7	−183.8	144.9	−6.2
$2NO_2 \rightarrow N_2O_4$	−12.7	−3.7	−182.8	144.9	−6.2

All values in [kJ/mol]

temperature for the water dimerization according to Eq. 3.14 which is in the order of magnitude of T^3 at $T = 298.15$ K ($T^3/\Delta\Theta^{2H_2O}_{(H_2O)_2} = 0.2$). The rotational temperature change mainly depends on the ratio of the principal moments of inertia of dimer and monomer (see Eq. 2.38). Due to the very small principal moments of inertia in the water monomer as compared to the dimer, this ratio is rather large and therefore the dimerization entropy is only affected very little by this contribution. The situation is different in the case of the hydrazine formation and especially in the case of the acetic acid dimerization. The difference in order of magnitude between the principal moments of inertia of the dimer and the monomer for the dimerization of NO_2 is considerably smaller and the change in rotational temperature (see Eq. 3.14) is largely affected by the quadratic term in the numerator ($A = B$ in the case of dimerization reactions).

Consequently, the argument of the logarithm function in Eq. 3.13 assumes considerably larger values as in the case of the water dimerization ($T^3/\Delta\Theta^{2NO_2}_{N_2O_4} \approx 8,700$). Following from that the change in rotational entropy is negative and of a considerably larger magnitude as the constant contribution (3/2)R as well as the value obtained for the rotational association entropy of water. The same reasoning is true for the association of acetic acid, but in that case the discussed effects are even more pronounced.

The partitioning of the translational association entropy summarized in the second block of Table 3.13 indicates large contributions at $T = 298.15$ K from the change in number density as well as the change in mass for all three model reactions, which are significantly larger in absolute value as compared to the constant contribution of (5/2)R. The magnitude of the system-unspecific $T\Delta S_{dens}$ value clearly arises due to an increase of the inverse particle density $(N_A/V)^{-1}$ (i.e., a decrease of the number density), thereby extending the entropy of the system. A reversed effect is observed for the translational entropy contribution due to the change in mass as expressed in Eq. 3.17.[6] For all investigated systems, an entropy loss of considerable magnitude is predicted, which in combination with the constant contribution of (5/2)R is responsible for the dominant effect of these degrees of freedom on the total entropy change as e.g. expressed in Table 3.11. The $T\Delta S_{mass}$ part is always larger in magnitude than the contribution arising from the change in number density, but the difference in mass between the various monomers in the three investigated systems is not as noticeable in the resulting entropy changes as in the case of the rotational degrees of freedom. This observation is based on the additional dependancy of the moments of inertia on the distance to the center of mass, which in general will be larger in the dimer as

---

[6] Please note that in an association reaction according to Eq. 3.3 there is of course no formal change in mass taking place. However, by calculating the association entropy as expressed in Eq. 3.12, the additive mass relation between the product and the reactants translates into a ratio according to Eq. 3.16, which is equal to the reduced mass of the system. This quantity relevant for the translational entropy change is not equal to one even if $m_A = m_B$ as in the case of dimerizations.

compared to the monomer. Thus, the rotational association entropy is not only affected by the increasing mass upon association, but also by the larger molecular dimension of the dimers, and these two effects finally lead to a more significant discrimination between the three examined systems in terms of the rotational entropy change. However, the translational association entropy is the most important contribution concerning the total magnitude, and the values listed in the second block of Table 3.13 show that this entropy change is a balance between two large opposing contributions due to the change in number density and a mass difference according to Eq. 3.16, with a predominance of the latter part.

Thus, the detailed analysis of the association entropy contributions from the interconversion of different degrees of freedom presented in this part supports the previously observed controlling influence of the translational degrees of freedom on the entropy change. The increase in vibrational entropy upon dimerization can clearly be attributed to the formation of six new low-frequency vibrations in the dimers rather than to an entropy production due to a shift of existing modes in the dimers. There is no evidence found for the compensating effect between the rotational and vibrational association entropy as observed for the pseudorotaxane association (see Table 3.7) neither in the numbers from Table 3.11 nor in the explicit expressions in Eqs. 3.13 and 3.10. However, the transition from the low density regime (for which the association entropies of the rrho approach are in good agreement to the experimental values, see Table 3.11) to the high density regime in the frame of entropy changes according to the rrho approach possibly has to be realized in terms of the translational degrees of freedom. The discussion so far clearly demonstrates that in reactions subject to a particle number effect as expressed in Eq. 3.3 the dominant contribution in all cases arises from the change in translational entropy. In addition, the results of the present part show that the translational association entropy (see Eq. 3.15) explicitly depends on the inverse of the number density according to the Sackur–Tetrode equation, which will be considerably different in the gas phase and in the condensed phase. Before proceeding on this subject, a short literature review of previous studies on the effect of changing particle numbers in entropy calculations will be given in the following.

3.2.1.3 The Particle Number Effect in Literature

The preceeding sections are mainly focussed on a discussion and quantification of the particle number effect occurring in association reactions of supramolecular pseudorotaxane complexes and some small molecule systems. However, early investigations concerning the subject of degree-of-freedom interconversion and the accompanying entropy change already took place in biochemical research during the 1950s and 1960s, when the thermodynamics of protein association were examined in greater detail. One of the first of these studies was carried out by Doty and Myers, who analyzed the equilibrium constant of the insulin monomer–dimer equilibrium through intensity reduction of scattered light [32, 33]. The measured reaction entropies of the dissociation reaction were found to lie between 0 J/(mol K) and up to 50 J/(mol K) averaged over a temperature range from 293 to 313 K depending on the pH value of the aqueous insulin solution. These results were in strong contrast to their theoretical estimate of about 122 J/(mol K) based on the gain of translational and rotational entropy upon dissociation of a structureless dimer in the gas phase. Some 9 years later Steinberg and Scheraga revisited the problem in a more rigorous fashion and pointed out that in protein association reactions where the association takes place via non-covalent interactions like hydrophobic bonding, the assumption of a complete loss of six translational and rotational degrees of freedom might be inaccurate [34]. Furthermore, they stressed the importance of a possible entropy recovering through new vibrational degrees of freedom in the associated product and gave a straightforward decomposition of the association entropy into contributions from the association itself (*intrinsic* entropy change) and the entropy change due to different solute-solvent interactions of the monomer and the dimer. In principle, this theoretical framework allows the calculation of the intrinsic entropy change according to the statistical thermodynamics of an ideal gas (i.e., in analogy to the rrho approach), whereas the solute-solvent contribution is expressed in terms of phase space integrals, for which the interaction potentials between the solute and solvent particles as well as between the solvent particles themselves have to be known. Due to these contributions an appropriate extension of the rrho approach can hardly be employed on the basis of static quantum chemical calculations. Through estimates of the entropy contributions arising from the different degrees of freedom they arrived at a total association entropy of −40 J/(mol K) for the dimerization of two structureless insulin particles in solution, a value which is in reasonable agreement to the measurements of Doty and Myers as well as to the numbers calculated for the pseudorotaxane exchange reaction, see Table 3.8. Although these results were referred to as not to be valid in the case of dimerizations involving more rigid covalent bonding, they clearly demonstrated the importance of a detailed treatment of the degree-of-freedom interconversion for the entropy difference in bimolecular reactions at an early stage.

The loss of translational and rotational entropy in bimolecular reactions was also recognized as an important factor in the discussion of the chelate effect and rate accelerations of intramolecular reactions [35, 36]. Page and Jencks pointed out that the rate acceleration in solution for an intramolecular reaction compared to the corresponding bimolecular reaction can be understood as an increased "effective molarity" of one reactant relative to the other, which is based on an increased number of rotational and translational entropy contributions to the reaction entropy [36]. It was assumed that due to a possible displacement of several solvent molecules upon the chelate ligand's association, there will be an additional gain in translational entropy and thus in effective concentration of the reactants concerning the overall reaction rate. However, these considerations are difficult to verify on the basis of common rrho calculations which rely on the single molecule picture of static quantum chemistry. At the concrete example of a Diels–Alder reaction the relevant aspects of the accurate treatment of solvent effects to the reaction entropy in solution as well as the residual entropy of the reaction product due to low lying vibrations were already addressed. One of the major conclusions of that study was the estimation of degree-of-freedom contributions to the entropy and the observation that in most cases the translational contribution exceeds those from rotations and vibrations, which is completely confirmed by the results presented in the previous parts of this thesis. In addition, the authors stated that even low-frequency harmonic oscillations at e.g. 100 cm^{-1} or internal rotations, for which estimates of S^0 = 14–19 J/(mol K) per mode were given, can barely compensate the translational contribution calculated to be between 121 and 151 J/(mol K) at molecular weights of 20–200 u and the standard concentration 1 M [36]. Again these numbers are in excellent agreement to the translational association entropies obtained in the previous sections on the basis of the rrho model, see Tables 3.7 and 3.11.

The interconversion of degrees of freedom in bimolecular association or dissociation reactions and the influence on the reaction entropy remained a subject of intensive research in the following years and gave rise to numerous publications mainly in the field of biochemical sciences, which covered issues like "enthalpy-entropy compensation" or "cratic entropy correction" [37–43]. One of the first more accurate treatments of residual entropy in the association product of an association reaction according to Eq. 3.3 in terms of semiquantitative calculations was given in 1992 by Doig and Williams, who investigated the thermodynamics of urea dimerization as a model system for the peptide bond in proteins [38]. These authors estimated the intrinsic entropy change of hydrogen bond formation by decomposition of the experimental reaction entropy of the dimerization reaction into (negative) contributions from the lost translational and rotational degrees of freedom and the replacement of the C–N internal rotation by a torsional harmonic oscillation. For the actual calculation of the single components a force field and the rrho model were employed in combination with a correction for the solvent effect, in which additional water molecules bound to the substrate and their effect upon mass and moments of inertia were considered as well as the condensation to the pure liquid and dilution to the 1 M reference state. Furthermore, the entropy gain due to newly formed vibrations in the reaction product was treated as part of the hydrogen bond formation entropy, which thereby corresponds to a true loss of six degrees of freedom in the dimer. In order to estimate these contributions, additional calculations of peptide monomers and dimers in non-polar solvents were carried out including a harmonic frequency analysis, and the authors found some new low-frequency vibrations in the dimer, for which an entropy contribution of approximately 60 J/(mol K) per hydrogen bond was predicted. One of their conclusions concerning new vibrations and solvent effects implied that in non-polar solvents as CCl_4 the newly introduced vibrations can account for nearly all of the hydrogen bond formation entropy and that their approach could not be easily transferred to more polar solvents due to transient bonds between solvent and solute. This is another example for a successful attempt to model entropic effects within a solvent by means of microsolvation as in the case of the exchange reaction for the pseudorotaxane association, but such approaches could be hardly feasible for routine standard quantum chemical calculations and the results are hardly transferable between different solvents.

Another molecular mechanics study concerning the entropy change in dimerization reactions was published by Tidor and Karplus in 1994 [41]. In this paper a refined approach to the dimerization of insulin was presented including a full normal mode analysis, thereby extending the structureless-particle model of Doty and Myers [32]. Furthermore, these authors applied a strict decomposition of the condensed phase reaction entropy into an intrinsic (gas-phase) contribution and the solvent contribution based on statistical mechanics in the manner of Steinberg and Scheraga, and stated that the use of gas-phase equations is a reasonable approximation if the solute–solvent interactions are constant over the conformational space sampled during the reaction course.

During the last decade, the application of the rrho approach in combination with static quantum chemical calculations has received considerable interest due to the increase in computational performance of hardware and software alike. Besides the association between supramolecular or biochemical compounds, reactions as expressed in Eq. 3.3 are found in the catalytic cycles of homogeneous transition metal catalysis. Numerous publications show that it is common practice to calculate changes in thermodynamic quantities for different steps of a catalytic cycle, which are employed for the prediction of the stabilities of intermediates and to obtain a deeper understanding of the underlying reaction mechanism [44–58]. Although it is often mentioned that calculated entropy changes are computed according to equations which are derived for gas phase reactions and that the corresponding values would be different in solution, only few studies apply correction schemes to account for that fact. Such corrections often rely on modified Born-Haber cycles in which the thermodynamic reaction quantities are calculated for the gas phase reaction and the transition to the condensed phase is carried out in some approximate way or is based on empirical observations concerning the entropy loss of compounds upon solvation [38, 59]. A prominent example of this kind is the observation of Wertz that all substances lose a comparable amount of their entropy upon transformation from the gas phase to aqueous solution and that during this process the entropy change of water does not significantly contribute to the solvation entropy of the substrate, even if considerable solute-solvent interactions like e.g. hydrogen bonds are formed [59]. The solvation entropy estimate ΔS_{solv} proposed in that study for the transition to aqueous solution is given by

$$\Delta S_{\mathrm{solv}} = -0.46S, \tag{3.18}$$

where S denotes the gas phase entropy of the substrate at a concentration of 55.5 mol/L. Estimates of this kind have been successfully combined with static quantum chemical computations for the entropy change occurring in ligand association or dissociation reactions by Zhu and Ziegler [56]. These authors applied a three-step process for the correction of the entropies calculated according to the rrho approach, in which the substrate is first compressed to the molar volume of the solvent, thereby changing its entropy by $\Delta S = R\ln(V_{m,\mathrm{f}}/V_{m,\mathrm{i}})$ where $V_{m,\mathrm{f}}$ and $V_{m,\mathrm{i}}$ denote the final and initial molar volume, respectively. Next, the entropy fraction α lost upon the solvation process is obtained from the absolute entropies of the pure solvent in its gas (S^0_{gas}) and liquid (S^0_{liq}) phase according to [56]

$$\alpha = \frac{S^0_{\mathrm{liq}} + (S^0_{\mathrm{gas}} + R\ln(V_{\mathrm{liq}}/V_{\mathrm{gas}}))}{S^0_{\mathrm{gas}} + R\ln(V_{\mathrm{liq}}/V_{\mathrm{gas}})}. \tag{3.19}$$

Finally, the substrate in solution is diluted to the 1 M reference state which involves an entropy contribution similar to the one of the first step. In the case of water, the authors confirmed the value for the fractional entropy loss of $\alpha = -0.46$ proposed by Wertz.

The approach of Zhu and Ziegler and the suggestion of Wertz demonstrate that the intrinsical shortcomings of the rrho approach concerning the particle number effect in entropy calculations can be corrected on the basis of empirical considerations. Numerous suggestions of this kind either depending on solvent-specific properties (as in the case of the approach by Zhu and Ziegler) [47, 56], on an estimate of the solvation entropy (as in the Wertz example) [57, 59], or on rules of thumb (as e.g. multiplying the gas phase reaction entropy by 2/3) [54, 55] can be found in the literature. However, no general strategy for the transition from the gas phase to the condensed phase with regard to the association entropy has been suggested on the basis of the conventional rrho approach, which does *not* depend on experimental measurements of solvent properties or phenomenological observations. The results of the previous part show that such a suggestion could be formulated most conveniently in terms of the translational association entropy, which explicitly depends on the inverse of the number density according to the Sackur–Tetrode equation (see Eq. 3.8). However, the contributions arising from the remaining degrees of freedom will also change upon transition from low to high densities, and it would be helpful for the design of a high density rrho approximation to permit a simplified treatment of these effects as well. Furthermore, a closer investigation of the approximations and the resulting errors occurring in the combined application of standard quantum chemical calculations and the rrho approach is of particular importance, because it is not directly clear how these errors affect the association entropy at different densities. The error analysis and the scaling parameters introduced in the next sections will provide a tool for the treatment of these factors.

3.2.2 Quantification of Errors

The previous section provides a detailed analysis of the particle number effect in reaction entropies of association reactions, and it is found that the association entropy of these reactions in the gas phase can be calculated to good precision, see Table 3.11. However, due to the small test set of investigated reactions the universality of this observation is not given, and the particle number effect in the case of the pseudorotaxane association reaction in solution gives rise to considerable differences between experimental measurement and the calculated values. The following parts will focus on a classification of the approximations and errors introduced in the general derivation of the rrho approach as presented in Sect. 2.1 and their quantification in terms of a numerical evaluation of the translational partition function as well as scaling factors to the mechanical properties relevant for the degree-of-freedom dependent partition functions. This first part aims at the classification of the different errors occurring in the combined application of the rrho model and static quantum chemical standard methods for the computation of the mechanical quantities required as input by the model. The employed approximations will be assigned to one of the three different classes:

- Model errors
- Model-inherent errors
- Technical errors

The class of model errors will feature the shortcomings arising from the primal assumptions of the model itself, i.e., approximations which are directly employed in the derivation of the rrho approach. These approximations largely contribute to the simplicity of the rrho model and enable the analytical formulation of this approach, thereby providing one of the most important advantages of this method as compared to many methods relying on molecular dynamics simulations, see Sect. 2.3. In addition, the molecular quantities required as input data to the model are readily obtained from static quantum chemical calculations, for which reason the rrho model can be applied as a natural and straightforward extension to such computations as can be seen in many actual quantum chemical program packages [60, 61]. However, this simplicity is not acquired without loss of accuracy due to the model assumptions, which in general are not fulfilled by the real system under investigation. Different from that class are errors arising *within* the applied methodology, i.e., errors which would be present even if the model would be exact. These are collected in the class of model-inherent errors and in the frame of the rrho approach mainly appear in the form of an inaccurate prediction of the mechanical quantities in terms of the applied quantum chemical methodology. Errors belonging to this class can be quantified in terms of scaling factors to the particular quantity, and analytical expressions for the entropy as a function of these scaling factors can be derived as will be shown in the following. The introduction of scaling factors is a common concept in quantum chemistry and can be efficiently employed for the correction of systematic errors due to an approximate methodology. For instance, it is well known that harmonic frequencies calculated in terms of the Hartree–Fock method overestimate the corresponding frequencies obtained from experiments, and therefore it is common practice to assign a scaling factor of approximately 0.9 to these values [62]. The last class of errors includes inaccuracies arising due to technical reasons. In the case of the computational rrho approach such errors are often of lesser importance as compared to the inaccuracies introduced by errors from the first two classes. In addition, these inaccuracies often can be reduced further in a systematic way, and their magnitude can be estimated to a good precision. The analysis of technical errors will not be covered in detail in this thesis, but a short qualitative summary of this class of inaccuracies will be given below.

3.2.2.1 Model Errors

The first class of errors arises due to the most obvious kind of approximations introduced by the design of the rrho model itself. In real molecular systems, rotations are not rigid due to centrifugal forces, oscillations are not harmonic due to deviations of the real potential energy surface from the parabolic form, and translational motion of molecular units does not take place in a potential-free cubic box, especially not at higher densities. However, the application of these highly ideal model systems is necessary for accessing the energy eigenvalues of the respective degrees of freedom as pointed out in Sect. 2.1.2 and for successfully establishing a connection between the rrho approach and results computable by electronic structure theory. For this reason and due to the lack of "exact" alternative formulations, a quantification of the errors introduced by the application of these model systems is very difficult and will not be covered in full detail in this thesis. One exception to this is given by the vibrational degrees of freedom, which can be systematically improved by extending the harmonic approximation in terms of anharmonic corrections. If such corrections are applied in the vibrational analysis, it is important to note that the harmonic oscillator partition function (Eq. 2.41) is also no longer applicable, since this equation is explicitly derived on the basis of the energy eigenvalues of a harmonic oscillator. Nevertheless, it is in principle possible to account for these corrections e.g. in terms of quartic potentials instead of the harmonic ones or by calculating anharmonic frequencies and the corresponding partition functions on the basis of vibrational self consistent field methods [63–65]. However, such approaches often come along with a considerable computational effort and are not yet established as standard methods for the calculation of thermodynamic properties of large or chemically complex systems. In addition, with regard to an evaluation of the rrho approach at liquid phase conditions the impact of anharmonic corrections on the vibrational entropies is expected to be rather small as e.g. compared to the effect of the number density on the translational entropy, which is found to be of considerable magnitude according to the results discussed in the previous sections (see e.g. Table 3.13). This reasoning is fully supported by a recent error analysis of the vibrational entropy according to the harmonic oscillator approach with regard to entropies obtained from a full vibrational configurational interaction (fvci) calculation for a single water molecule [65]. The results of that study demonstrate that the entropies calculated from the harmonic oscillator model and the fvci approach differ by approximately $T\Delta S = 0.01$ kJ/mol at a temperature of $T = 298.15$ K, which is far less than the inaccuracies in entropy expected from most other approximations of the rrho model.[7] Thus, the analysis of model errors will focus on a quantification of contributions arising from the treatment of the translational degrees of freedom in the rrho model instead of the treatment of rotations as being rigid and vibrations as being harmonic.

Besides these approximations, the derivation of the rrho approach in Sect. 2.1.2 involves additional inaccuracies which can be classified as model errors. As pointed out in detail in Sect. 2.1.1, the factorization of the canonical N-particle partition function Q into the single particle contributions q relies on the indistinguishable and (more important) ideal character of the constituting particles, thereby ignoring any form of spatial dimensionality and interaction between the particles. The error introduced by this ideal gas approximation is clearly more severe for the treatment of condensed phase processes, but even in the gas phase the importance of interparticle interactions cannot generally be ignored. For instance, the gas phase of hydrogen fluoride at ambient conditions contains a significant number of six-membered rings which would not be stable if the intermolecular hydrogen bond is neglected [66, 67]. Thus, the isolated molecule picture applied in static single molecule calculations (and in the factorization of the rrho partition function) is often different from the real gas phase, and one particular way of considering interparticle interactions (which will be investigated in the next chapter) are cluster approaches treating more than a single particle in the quantum chemical calculation. The single particle partition functions q are generally still too complicated to calculate for most molecular systems, which is the reason for an additional factorization of the single particle partition function into its contributions from the molecular degrees of freedom as laid out in detail in Sect. 2.1.2. This second factorization implies that each degree of freedom is independent from all others and can be treated in terms of a separate partition function. However, it is known from experiment as well as theory that certain molecular degrees of freedom do couple, as can be seen in the finite values of, e.g., rotation-vibration coupling constants [68, 69]. The magnitude of these coupling constants can provide information about the accuracy of the factorization into different degrees of freedom (and whether it is reasonable at all), but without knowledge of such system-specific properties the extent of this approximation and the resulting error can hardly be estimated in a general way and with regard to the density of the system.

Besides the factorizations of the partition functions Q and q, additional model errors arise in the evaluation of the translational partition function q_{trans} and the rotational partition function q_{rot} due to the approximate treatment of the summation as an integral, see Eqs. 2.30 and 2.35. These model errors are accessible to a numerical evaluation and will be quantified in the following for the translational

[7] It should be noted that these small differences also arise due to the relatively large wavenumbers of the vibrations in the water monomer, which affect the vibrational entropy only to a small extent (see Eq. 3.10). Nevertheless, deviations in the order of magnitude of only some J/(mol K) can be estimated for the low-frequency modes occurring in the dimers as well, which justifies a preferential treatment of the remaining degrees of freedom.

entropy of a water molecule at $T = 298.15$ K, thereby following the observation that the translational entropy contributions show the largest impact on the total association entropies, see Tables 3.7 and 3.11.

The accuracy of the high temperature approximation for the translational degrees of freedom can be estimated by calculating the difference in the argument of the exponential function $\beta\Delta\epsilon_{\text{trans}}$ in Eq. 2.29 for two adjacent values of l, which is given by [21]

$$\beta\Delta\epsilon_{\text{trans}} = \beta\left(\frac{h^2(l+1)^2}{8ma^2} - \frac{h^2l^2}{8ma^2}\right) = \beta\frac{h^2(2l+1)}{8ma^2}. \tag{3.20}$$

From this equation it is apparent that the difference between the exponential terms in Eq. 2.29 decreases as the temperature is increased which is to be expected according to the high temperature limit. For a water molecule of mass $m = 18$ u and a box length of $a = 10$ cm, the translational quantum number l assumes values in the order of magnitude of 10^9–10^{10} at room temperature, and the difference in translational energy states weighted by temperature according to Eq. 3.20 amounts to approximately 3×10^{-10} [21]. Consequently, the effect of the continuum approximation to the translational energy spectrum can be expected to be rather small at these conditions. However, a box length of 10 cm is not a realistic estimate for the range available for translational motion in the condensed phase. The length scale of translational motion for a water molecule in the liquid phase at ambient conditions can be estimated from the density and the molar weight of water to be in the order of magnitude of 3×10^{-10} m or even smaller (a more rigorous approach to the estimation of the free translational volume in the condensed phase will be presented at the end of the chapter). At these conditions a typical value for the translational quantum number is given by $l = 10$ according to Eq. 2.28 and the equipartition theorem. Accordingly, the thermally weighted difference expressed in Eq. 3.20 assumes values as large as $\beta\Delta\epsilon_{\text{trans}} = 0.1$, which clearly is in a different order of magnitude as the value obtained for the previous conditions. Thus, the distance between adjacent translational energy states amounts to approximately 10% of the available thermal energy β^{-1}, and the integral approximation of the sum in the translational partition function according to Eq. 2.30 can be expected to be less accurate as in the case of macroscopic box dimensions. In order to quantify the accuracy of the high temperature limit for the translational partition function in the condensed phase, values obtained from the exact equality as expressed in Eq. 2.29 have to be compared to numbers obtained from the integral approximation in Eq. 2.30 at translational length scales representative for liquid phase densities. The exact evaluation of the sum in Eq. 2.29 is in principle not possible (which is the reason for the application of the integral approximation) due to the infinite number of terms, but the inverse quadratic dependancy of each summand on the translational quantum number l should lead to an acceptable convergence behavior, and the estimate for l according to the equipartition theorem indicates that relatively small values of the translational quantum number yield the most important contributions to the sum in Eq. 2.29 at

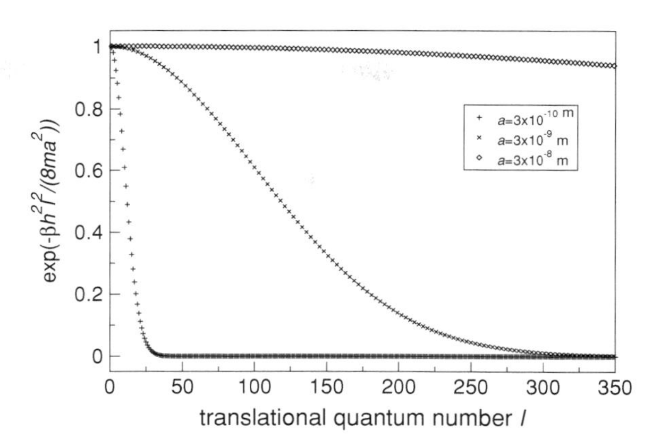

Fig. 3.4 Distribution of terms appearing in the translational partition function (Eq. 2.29) with varying translational quantum number l for three different box lengths ($a = 3 \times 10^{-10}$ m, $a = 3 \times 10^{-9}$ m, and $a = 3 \times 10^{-8}$ m) at $T = 298.15$ K. For the sake of clarity only every second term is plotted in the two latter cases

length scales typical for the condensed phase. This can clearly be seen from the distributions of the individual terms as a function of the translational quantum number l as illustrated in Fig. 3.4. For a translational scale of approximately $a = 3 \times 10^{-10}$ m estimated for liquid water at ambient conditions, a steep decline of the translational exponential terms is observed and the only significant contributions arise from translational quantum numbers of up to $l = 30$. Within this interval, the value of these terms decreases by two orders of magnitude, and from a numerical point of view the zero limit for $l \to \infty$ is virtually reached for $l \approx 65$.

By restricting the evaluation of the translational partition function according to Eq. 2.29 to the first 30 terms in the sum, 99% of the value obtained for q_{trans} on the basis of the first 2,000 terms is recovered. These considerations demonstrate that at liquid phase densities a numerical evaluation of the translational partition function is feasible and that such a calculation could be a more accurate alternative as compared to the integral approximation. The situation is different if the translational scale a is raised by a factor of ten. In this case the decline of terms obtained from larger values of l is not as pronounced as before, and considerable contributions are found for translational quantum numbers of up to $l = 300$. The zero limit for $l \to \infty$ is numerically obtained for quantum numbers as large as $l = 650$, and the first 30 terms of the sum in Eq. 2.29 only contribute approximately 1% to the value of q_{trans} calculated from the first 2,000 terms. Thus, an accurate evaluation of q_{trans} on the basis of the exact equality in Eq. 2.29 becomes increasingly difficult if the translational scale is enlarged. This is even more obvious if the box length is again raised by a factor of ten to a value of $a = 3 \times 10^{-8}$ m. Now the first 30 terms yield a constant contribution of 1.0 according to the limit $l \to 0$, i.e., the fraction l^2/a^2 is so small that this limit is numerically reached for the first 30 terms, and the decline of the distribution is considerably slower as for the previously examined cases. This can clearly be seen in the large contribution arising from the term for $l = 2{,}000$, which is equal to 1.38×10^{-1} and thereby amounts to approximately 14% of the first term in the sum. From these observations it is clear that for length scales in the order of magnitude of 10^{-8} m and larger the sum in the exact equality for q_{trans} does not converge within the first few thousand terms and that a numerical evaluation of the translational partition function represents a considerable task which could be too laborious for standard applications. However, at translational volumes in the order of magnitude estimated to be relevant for liquid water, an acceptable convergence of the sum in Eq. 2.29 can be expected, and on the basis of this observation a comparison between the exact equality and the integral approximation for the translational partition function can be carried out at these conditions.

However, there is still one important detail which has to be considered in order to perform a reasonable comparison between the numbers obtained from Eqs. 2.29 to 2.30. The lower limit of the integral in Eq. 2.30 is equal to zero, i.e., the lowest value the integration variable l can assume is zero, thereby allowing a convenient evaluation of the Gaussian-type integral. In contrast, the explicit summation over the Boltzmann factors of the particle-in-a-box energy eigenstates is accomplished in terms of the quantum number l, which can assume values from the set of positive natural numbers, i.e., a term for $l = 0$ does not occur in Eq. 2.29 according to the quantum mechanical treatment of the particle-in-a-box problem (for $l = 0$ the wave function of the particle would also be equal to zero and therefore no normalization would be possible). This is no problem in the high temperature limit where l is treated as a continuous variable according to the assumption of a continuum of energy states, since in this approximation l can assume any value arbitrarily close to zero. In order to account for this difference in the energy scale between both approaches for the calculation of q_{trans}, one has to ensure that the lowest possible energy state obtained for $l = 1$ is equal to zero, which formally corresponds to a translational quantum number of zero according to Eq. 2.28. This can be done most directly by the introduction of a shifted spectrum of eigenstates $\{\epsilon^{\dagger}(l)\}$, which differs from the original spectrum by the energy of the ground state according to

$$\epsilon^{\dagger}(l) = \epsilon(l) - \epsilon(1), \tag{3.21}$$

i.e., the energy of the lowest state in the shifted spectrum is equal to zero. The effect of these shifted translational energy eigenvalues on the translational partition function $q^{\dagger}_{\text{trans}}$ is directly obtained from Eqs. 2.29 and 3.21 according to

$$\begin{aligned} q^{\dagger}_{\text{trans}} &= \left[\sum_{l}^{\infty} \exp(-\beta\epsilon^{\dagger}(l))\right]^3 = \left[\sum_{l}^{\infty} \exp(-\beta[\epsilon(l) - \epsilon(1)])\right]^3 \\ &= \left[\sum_{l}^{\infty} \exp(-\beta\epsilon(l)) \times \exp(\beta\epsilon(1))\right]^3 = \left[\exp(\beta\epsilon(1)) \sum_{l}^{\infty} \exp(-\beta\epsilon(l))\right]^3 \\ &= \exp(3\beta\epsilon(1)) q_{\text{trans}}, \end{aligned} \tag{3.22}$$

i.e., the partition function is affected by a constant factor of $\exp(3\beta\epsilon(1))$. The comparison between q_{trans} obtained from the high temperature limit as well as calculated according to Eq. 3.22 is illustrated in Fig. 3.5, in which the natural

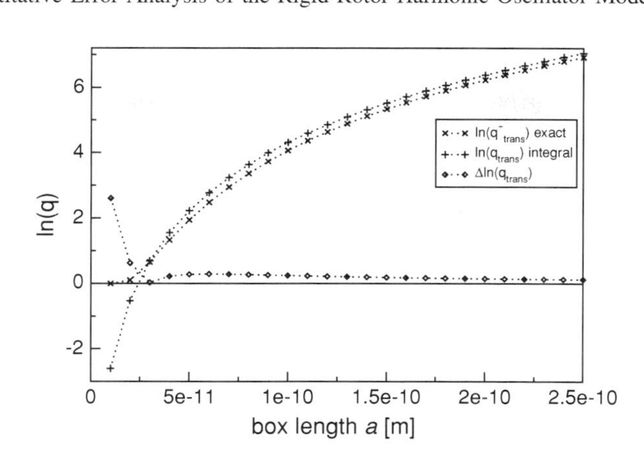

Fig. 3.5 Natural logarithm of the translational partition function calculated according to the shifted summation (Eqs. 3.22 and 2.29) and according to the integral approximation (Eq. 2.30) as a function of the box length a at $T = 298.15$ K. $\Delta \ln(q_{\text{trans}})$ gives the magnitude of difference between the exact result and the one obtained from Eq. 2.30. For the evaluation of Eq. 2.29, a constant upper summation limit of $l = 2{,}000$ is chosen

logarithm of the translational partition function obtained from these two different approaches is plotted as a function of the box length. In this comparison the natural logarithm is employed instead of the actual value of the partition function itself due to the many order of magnitudes which q_{trans} adopts in this length scale interval. Furthermore, the natural logarithm is the relevant quantity on which the entropy depends on in the (N, V, T) ensemble according to Eq. 2.64. The plots in Fig. 3.5 demonstrate that at length scales of $a = 2.5 \times 10^{-10}$ m the replacement of the sum by an integral according to Eq. 2.30 is a rather accurate approximation at $T = 298.15$ K. There is only a small difference between the logarithms of q_{trans} calculated from Eqs. 3.22 to 2.30, respectively, as can be seen from the difference function $\Delta \ln(q_{\text{trans}})$ which assumes values very close to zero in this region. Increasing deviations between the two approaches are found at length scales of $a = 1.5 \times 10^{-10}$ m and smaller. In this region, the result obtained from Eq. 3.22 constantly yields values of lower magnitude as compared to the ones calculated according to the integral approximation, and the differences are increasing with a further decrease of the box length as can be seen in the values of the difference function $\Delta \ln(q_{\text{trans}})$ deviating from zero. At translational scales of $a = 5 \times 10^{-11}$ m and smaller the difference in q_{trans} calculated from the two approaches decreases again, and at approximately $a = 3 \times 10^{-11}$ m both curves cross each other. However, beyond that point both partition functions diverge rather fast, which is the reason for a steep rise of the difference function $\Delta \ln(q_{\text{trans}})$ at this length scale and smaller values of a. The curve obtained from the numerical evaluation (Eq. 3.22) declines less significantly than before and asymptotically approaches zero in the limit $a \to 0$. This behavior is to be expected, since smaller box lengths formally correspond to a decrease of temperature according to

Eq. 2.29, and the canonical partition function approaches one in the limit of $T \rightarrow 0$ due to the vanishing contribution from the higher terms in the sum of Eq. 3.22 [70].

However, this progression is not observed for the curve calculated from the high temperature limit, which is of the functional form $f(x) = \ln(x^{3/2})$ according to Eq. 2.30 and thereby represents the graph of an analytic logarithm function. Consequently, in the limit of $a \rightarrow 0$ the integral approximation for the calculation of q_{trans} behaves like the ordinary natural logarithm function and approaches $-\infty$. This contrary trend observed for the two approaches leads to significant differences at the lower end of the examined length interval. Below $a = 2.5 \times 10^{-11}$ m the difference function deviates from zero by a considerable amount, and at the smallest box length considered ($a = 1 \times 10^{-11}$ m) the result obtained from Eq. 3.22 differs from the one according to the integral approximation by a factor of approximately 14. This clearly indicates that at translational scales of roughly $a = 1 \times 10^{-10}$ m or lower the high temperature approximation to q_{trans} as expressed in Eq. 2.30 might introduce a non-negligible error to the translational entropy, but at the translational scale of approximately $a = 3 \times 10^{-10}$ m estimated previously for the case of liquid water there is only a very small difference between the two approaches. Hence, the error introduced to the translational entropy is negligible at these conditions. However, this value of a is only a rough order-of-magnitude estimate of the translational length scale in liquid water and for instance neglects the distinctive hydrogen bond network found in the liquid water phase, which will additionally constrain the translational motion of individual water molecules [71, 72]. Therefore, the actual translational length scale in liquid water could as well be smaller than $a = 3 \times 10^{-10}$ m, but since substantial large deviations between the two investigated approaches do not show up at length scales above $a = 2.5 \times 10^{-11}$ m (which is equal to approximately one quarter of the oxygen–hydrogen bond length in the water molecule), the effect on the translational entropy at ordinary liquid phase densities can be expected to be rather small.

In order to confirm these considerations, the difference in translational entropy calculated from the shifted sum in Eq. 3.22 and from the high temperature limit will be quantified in the following. As laid out in the discussion of the Sackur–Tetrode equation (see Eq. 3.8), the factorial term $N!^{-1}$ in the canonical partition function (see Eq. 2.20) is usually combined with the translational entropy and evaluated according to Stirling's approximation. However, at translational length scales as small as those illustrated in Fig. 3.5, this contribution is considerably larger ($-R\ln(N_A) + R = -133.3\,\text{kJ/mol}$ at $T = 298.15$ K) than the true translational entropy obtained from q_{trans}, which will result in a shift of the translational entropy to negative values due to the negative sign of the $-R\ln(N_A)$ term. For this reason only the true translational contribution as expressed in Eq. 3.7 will be analyzed. In the case of the high temperature limit, this procedure leads to a modified Sackur–Tetrode equation for the translational entropy, which is given by

$$S_{\text{trans}} = R\ln\left(V\left[\frac{2\pi mk_{\text{B}}T}{h^2}\right]^{3/2}\right) + \frac{3}{2}R, \qquad (3.23)$$

i.e., all dependencies on the particle number are eliminated from the equation. In analogy, only the first term in the third line of Eq. 3.7 will be considered in the case of the numerical evaluation of q_{trans}. According to the definition in Eq. 2.64, the derivative of the logarithmized partition function with respect to the temperature has to be calculated in addition to the absolute value, which is already plotted in Fig. 3.5. In the case of the modified Sackur–Tetrode equation, this derivative term yields a constant contribution of (3/2)R as can be seen in Eq. 3.23. The situation is different for the shifted partition function in Eq. 3.22, since each single term of the sum depends exponentially on the inverse temperature and the shift of the translational energy states according to Eq. 3.21 has to be considered. For that case the differentiation with respect to temperature is thus given by

$$\begin{aligned}\left(\frac{\partial \ln(q^{\dagger}_{\mathrm{trans}})}{\partial T}\right)_{N,V} &= \frac{\partial}{\partial T}\ln\left(\exp(3\beta\epsilon(1)) \times \left[\sum_{l}^{\infty}\exp(-\beta\epsilon(l))\right]^{3}\right) \\ &= 3\left[\left(\sum_{l}^{\infty}\exp(-\beta\epsilon(l))\right)^{-1} \times \sum_{l}^{\infty}\frac{\partial}{\partial T}\exp(-\beta\epsilon(l))\right] + 3\frac{\partial}{\partial T}\beta\epsilon(1) \\ &= 3\left[\left(\sum_{l}^{\infty}\exp(-\beta\epsilon(l))\right)^{-1} \times \sum_{l}^{\infty}\frac{\beta\epsilon(l)}{T}\exp(-\beta\epsilon(l))\right] \\ &\quad - 3\frac{\beta\epsilon(1)}{T}. \end{aligned} \tag{3.24}$$

The translational entropies calculated from these equations are illustrated in Fig. 3.6 as a function of the translational length scale. For the numerical approach the contribution to the entropy arising from the absolute term $R\ln(q^{\dagger}_{\mathrm{trans}})$ and the derivative term $RT(\partial \ln(q^{\dagger}_{\mathrm{trans}})/\partial T)$ are plotted separately. It is apparent that in analogy to the curves in Fig. 3.5 there is only a very small difference in translational entropy calculated according to the high temperature limit and from the partition function $q^{\dagger}_{\mathrm{trans}}$ down to length scales of approximately $a = 1 \times 10^{-10}$ m. Below this distance, the two curves start to show larger differences, and at length scales of approximately $a = 2.5 \times 10^{-11}$ m both approaches predict a completely different entropy trend. At these distances, the translational entropy obtained from the high temperature limit declines very fast and crosses the zero entropy line at approximately $a = 1.5 \times 10^{-11}$ m, and at even smaller length scales a negative translational entropy is predicted by this approximation with $S_{\mathrm{trans}} \rightarrow -\infty$ in the limit $a \rightarrow 0$. Such a trend is physically unreasonable, since the smallest entropy a system can assume is zero according to the third law of thermodynamics, but is to be expected due to the logarithmic dependancy of the translational entropy on the volume as expressed in Eq. 3.23 [21].

Thus, the integral approximation can be classified as being accurate at length scales of approximately $a = 1 \times 10^{-10}$ m and above, but smaller translational

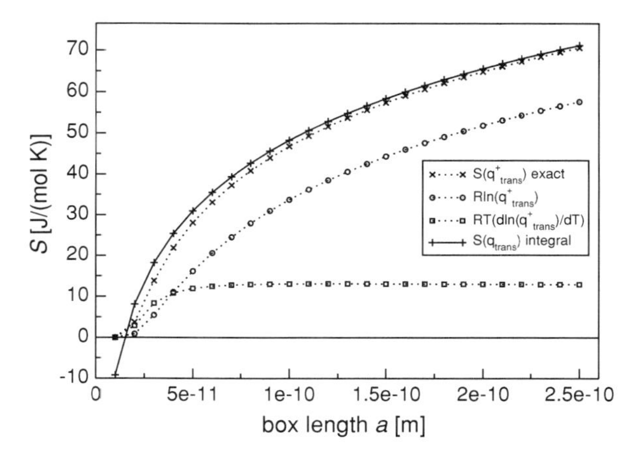

Fig. 3.6 Translational entropies calculated according to the shifted summation (Eqs. 3.22 and 2.29) and to the integral approximation (Eq. 2.30) as a function of the box length a at $T = 298.15$ K. $R\ln(q^{\dagger}_{\text{trans}})$ and $RT(\partial\ln(q^{\dagger}_{\text{trans}})/\partial T)$ denote the individual contributions from the absolute and the derivative term according to Eq. 2.64 for the numerical evaluation of $q^{\dagger}_{\text{trans}}$. For the evaluation of Eq. 2.29, a constant upper summation limit of $l = 2{,}000$ is chosen

scales lead to increasing inaccuracies and a physically not reasonable entropy prediction below a box length of approximately $a = 1.5 \times 10^{-11}$ m. In contrast, the translational entropy calculated from the numerical evaluation of Eq. 3.22 assumes constantly smaller values of approximately 5 J/(mol K) than those calculated from the high temperature limit at length scales below $a = 1 \times 10^{-10}$ m. At even lower length scales, the slope of this curve decreases significantly and no longer parallels the one predicted by the high temperature approximation. In the limit $a \to 0$ the translational entropy obtained by the numerical evaluation of $q^{\dagger}_{\text{trans}}$ approaches $S_{\text{trans}} \to 0$, which is in complete contrast to the high temperature prediction in the corresponding limit. However, this is a physically reasonable progression which is to be expected according to the third law of thermodynamics, since a variation of the translational scale is qualitatively equivalent to a variation in temperature, see Eq. 2.29. From the curves calculated for the individual contributions $R\ln(q^{\dagger}_{\text{trans}})$ and $RT(\partial\ln(q^{\dagger}_{\text{trans}})/\partial T)$ it is apparent that at length scales larger than $a = 5 \times 10^{-10}$ m the translational entropy is largely determined by the absolute term $R\ln(q^{\dagger}_{\text{trans}})$ concerning magnitude and especially the slope of the curve, whereas the derivative term $RT(\partial\ln(q^{\dagger}_{\text{trans}})/\partial T)$ yields a constant contribution of approximately 10 J/(mol K). Both contributions cross at approximately $a = 4 \times 10^{-11}$ m, and at even smaller distances the derivative term starts to decline as well, but not as fast as the absolute term. Consequently, the largest contributions to the entropy in this length domain arise from the derivative term, but the estimated differences of about 2–3 J/(mol K) are rather small. In the limit of very short distances, both terms approach zero, which is the reason for the translational entropy to behave equally in the numerical evaluation scheme of the translational partition function. Thus, the translational entropies predicted by the integral approximation and the numerical summation of the translational partition function are almost identical over a very large translational scale. Deviations between the two approaches start to show up at length scales of approximately $a = 1 \times 10^{-10}$ m and below, thereby demonstrating that the high temperature limit is an accurate approximation at gas phase densities and possibly also at ordinary liquid phase densities. Care has to be taken at very short translational scales of approximately $a = 5 \times 10^{-11}$ m and below, at which both predictions show larger discrepancies. This is especially apparent in the limit $a \rightarrow 0$, in which the high temperature approximation approaches a physically not reasonable translational entropy of $S_{\mathrm{trans}} \rightarrow -\infty$. However, the significance of this behavior at length scales considerably below those of most chemical bonds for translational entropies in the condensed phase is questionable.

In addition to the application of the high temperature limit to the translational and rotational partition functions, Stirling's approximation is typically invoked for standard entropy calculations in the frame of the rrho model in order to combine the factorial term of the canonical partition function in Eq. 2.26 with the translational entropy as expressed in the Sackur–Tetrode equation, see Eq. 3.8. Stirling's approximation can be understood as an integral approximation and therefore belongs to the class of model errors in complete analogy to the high temperature limit, since the sum appearing in the logarithm of the factorial expression

$$\ln(N!) = \sum_{n=1}^{N} \ln(n) \tag{3.25}$$

is approximated by an integration according to

$$\ln(N!) = \sum_{n=1}^{N} \ln(n) \approx \int_{1}^{N} \ln(n) dn = N \ln(N) - N \tag{3.26}$$

for large values of N [21]. In addition, Stirling's approximation is an example for an asymptotic approximation to an analytic function, which implies that it will become more accurate as the magnitude of the function argument is increased [21]. In analogy to the procedure applied in the case of the high temperature limit, the accuracy of Stirling's approximation is most directly quantified by comparing the results of the exact equation to the one obtained from the approximate expression. Such a comparison for increasing values of N is summarized in Table 3.14 [22]. The numbers listed in Table 3.14 indicate that for $N > 1{,}000$ Stirling's approximation can be seen as being exact with regard to the entropy calculations in the frame of the rrho model, because in that model Stirling's approximation is usually applied to resolve the term $k_{\mathrm{B}} \ln(N!)$, and in almost all cases N is in the order of magnitude of Avogadro's number N_A. For these values of N, no difference between the exact expression and Stirling's approximation can be computed via standard

Table 3.14 Results obtained from Stirling's approximation $N \ln(N) - N$ and the exact expression $\ln(N!)$ for increasing values of N

N	$\ln(N!)$	$N \ln(N) - N$	Deviation [%]
1×10^0	0.00	−1.00	–
1×10^1	1.51×10^1	1.30×10^1	13.8
1×10^2	3.64×10^2	3.61×10^2	0.9
1×10^3	5.91×10^3	5.91×10^3	0.1
1×10^4	8.21×10^4	8.21×10^4	0.0
1×10^5	1.05×10^6	1.05×10^6	0.0

The last column gives the deviation between both results in percent [22]

applications, which demonstrates that the effect of this approximation on the calculated entropies can be neglected.

Hence, there are three approximations classifiable as model errors in the rrho approach, namely the decoupling of the N-particle system (and of different molecular degrees of freedom), Stirling's approximation, and the approximation introduced by the quantum model systems including integral substitution of sums over energy states. The first of these model errors is based on the complete neglect of intermolecular interactions and will be examined in detail in the following chapter in terms of cluster approaches to condensed phase thermodynamics. The error arising from the application of the quantum mechanical model systems is difficult to quantify in the case of the particle-in-a-box model and the model of the rigid rotor, because a more elaborate treatment of the corresponding degrees of freedom soon leads to models too complicated to be routinely employed in standard applications [21, 73]. In the case of the harmonic approximation, a quantification is possible in terms of anharmonic corrections to frequencies and the vibrational partition function, but their effect on the entropy are calculated to be small for the case of water at ambient conditions by the fvci method [65]. However, the situation can be different if a considerable number of low-frequency vibrations is present, which are often poorly treated by a harmonic potential but at the same time yield the largest entropy contributions, see Table 3.12 [37, 74]. This situation can e.g. occur if internal rotations within the molecule are treated as harmonic vibrations. However, a numerical analysis carried out for the gas phase association between hydrogen fluoride and chlorine monofluoride indicates that the rrho approximation is again applicable with a reasonable accuracy for entropy calculations at medium temperatures like e.g. room temperature [75, 76]. The analogous numerical analysis for the assessment of the integral approximation to the translational entropy contribution and the application of Stirling's approximation presented in this section demonstrates that these integral approximations are accurate over a rather large interval of the relevant parameters as well. The numerical result for the translational entropy differs from that of the high temperature approximation considerably only at small translational scales of approximately one hundred picometers and below, and Stirling's approximation can be classified as being accurate for those numbers typically occurring in the treatment of chemical systems.

At the end of this part it should be noted that in the usual combination of static quantum chemical calculations and the rrho approach another form of model error is often encountered, which results from the neglect of conformational effects at finite temperatures. At a given temperature several different conformers will typically make a relevant contribution to the thermodynamic state in most systems according to their Boltzmann factor, but in many quantum chemical applications these contributions are neglected and only one minimum structure is considered. The effect of this model error is highly system-specific though, and for a given system and a given temperature could be estimated in terms of md simulations in the (N, V, T) ensemble (or an appropriate Monte Carlo sampling). However, this chapter concentrates on the calculation of thermodynamic quantities from the single molecule approach of static quantum chemistry and its link to the rrho model via the mechanical properties obtained from the electronic structure calculation, and the effect of inaccuracies in these properties will be quantified in the following section.

3.2.2.2 Model-Inherent Errors

The results and the discussion presented in the previous part is focussed on the errors in entropy arising from the approximations which are typically invoked during the derivation of the rrho model, see Sect. 2.1.2. In contrast, the following content will concentrate on a quantification of errors in entropy predictions due to inaccuracies *beyond* the ones which are introduced by the model itself, i.e., errors occuring *within* the applied methodology. According to the formulation of the rrho approach as laid out in Sect. 2.1.2, the molecular properties which have to be available for the evaluation of the molecular partition function are the molecular mass m and the principal moments of inertia $I_{a,b,c}$ for the translational and rotational degrees of freedom, respectively, and the full set of harmonic frequencies $\{\nu_k\}$ in the case of the vibrational partition function. In addition, the degeneracy of the electronic ground state has to be known in order to calculate the contribution from that state to the electronic partition function, and if excited electronic states are thermally accessible the respective energies also have to be provided for a numerical evaluation of q_{el} according to Eq. 2.43. The determination of these quantities and the effect of possible inaccuracies in that determination is no direct consequence of the rrho approach, and therefore these errors are separated from the model errors in the class of model-inherent errors. A possible way for the determination of the above listed molecular properties are experimental measurements, but the analysis of this section will concentrate on the calculation via static quantum chemical methods. However, the quantification of errors in entropy predictions presented below can also be applied to inaccuracies due to experimental measurements.

The most general approach for the computational determination of the required molecular properties is the execution of a structure optimization of the system under investigation followed by a normal mode analysis (see Sect. 7.1).

The optimized geometry will provide information about the principal moments of inertia, whereas the normal mode analysis yields the full set of harmonic frequencies. The determination of the molecular mass is a straightforward procedure for which no electronic structure calculation is required. It should be noted that isotopic effects are rarely considered in the structure optimization and the normal mode analysis, i.e., the mass of individual isotopes has to be applied in the determination of the molecular mass rather than the atomic weights. However, translational entropies according to the rrho approach are barely affected by these differences, e.g. the difference in translational entropy for a monomer of heavy water and one monomer of normal water is less than 2 J/(mol K) at $T = 298.15$ K. The situation will be more complicated in the case of the principal moments of inertia and the harmonic frequencies, because these quantities directly depend on the quantum chemical methodology and thereby are normally exposed to several approximations from the chosen method. On the level of the quantum chemical methodology most of these approximations can either be classified as model errors (e.g. the neglect of excited determinants in single reference methods) or technical errors (e.g. finite values for the convergency criteria of an iterative calculation), but in the frame of the rrho approach these inaccuracies are labeled as model-inherent errors, because they do not arise within the rrho model itself. A direct quantification of such model-inherent errors on the theoretical basis of the chosen electronic structure method is very complicated due to the interference of technical errors which are always present in the computational implementation of a given methodology. In addition, an incorporation of these expressions into the evaluation of the partition functions for the rotational and vibrational degrees of freedom according to the procedures introduced in Sect. 2.1.2 would have to be accomplished.

The approach pursued in this section will be different insofar as the individual sources of inaccuracies arising from the application of a certain electronic structure method are merged into a universal scaling factor δ_x for each of the relevant molecular properties ($x \in \{m, \rho_n^{-1}, I_{a,b,c}, \{\nu_k\}\}$) (with regard to a high density assessment, both the mass m as well as the inverse number density ρ_n^{-1} will be subject to the scaling procedure even though these quantities are not obtained from the electronic structure calculation). There are extensive studies available in the literature evaluating the performance of many different electronic structure methods with regard to structure optimization and normal mode analysis [62]. From these investigations the uncertainty of the applied methodology for the calculation of the particular property x can be estimated and translated into an appropriate scaling factor, which can subsequently be employed for the determination of the corresponding error in the entropy δS_y according to

$$\delta S_y(\delta_x, x) = S_y^{\text{ideal}}(x) - S_y^{\text{real}}(\delta_x, x), \tag{3.27}$$

where $S_y^{\text{ideal}}(x)$ denotes the unscaled entropy contribution from the regular rrho model for the particular degree of freedom y (y: trans, rot, vib) and $S_y^{\text{real}}(\delta_x, x)$ the entropy being subject to inaccuracies in terms of the corresponding scaling factor δ_x.

The first application of this scheme will be the scaling of the translational entropy as a function of the mass m and the inverse number density ρ_n^{-1}. According to the analytic form of the Sackur–Tetrode equation (Eq. 3.8) the corresponding error functions $\delta S_{\text{trans}}(\delta_m)$, $\delta S_{\text{trans}}(\delta_{\rho_n^{-1}})$ only depend on the scaling factors δ_m, $\delta_{\rho_n^{-1}}$ and are given by the following expressions

$$\delta S_{\text{trans}}(\delta_m) = \frac{3}{2} R \ln\left(\frac{1}{\delta_m}\right) \tag{3.28}$$

$$\delta S_{\text{trans}}(\delta_{\rho_n^{-1}}) = R \ln\left(\frac{1}{\delta_{\rho_n^{-1}}}\right). \tag{3.29}$$

From the inverse logarithmic dependancies expressed in Eqs. 3.28 and 3.29 it is clear that the error will be zero for a unit scaling $\delta_m = \delta_{\rho_n^{-1}} = 1$ and that the error will increase as the scaling factors deviate from unity. A plot of the translational error functions is illustrated in Fig. 3.7. From these curves it is apparent that non-negligible errors of $\delta S_{\text{trans}} = 5$ J/(mol K) in the translational entropy are found for scaling factors of approximately $\delta_{\rho_n^{-1}} = 0.6$ and $\delta_m = 0.8$. There is only a moderate increase in both error functions over a rather wide interval of scaling factors, with $\delta S_{\text{trans}}(\delta_m)$ rising faster at smaller values of the corresponding scaling factor as it is to be expected according to Eq. 3.28. At scaling factors of $\delta = 0.2$ and below the increase of the error in the translational entropy is more pronounced already for small variations in δ, and in the limit $\delta \rightarrow 0$ both curves approach $\delta S_{\text{trans}} = \infty$ which again is apparent from Eqs. 3.28 to 3.29. However, it is also obvious that substantially large errors in the translational entropy are only obtained in the case of very small scaling factors. Even if the relevant property is reduced to 10% of its original value (i.e., a scaling parameter of $\delta = 0.1$ is applied), the error in translational entropy only amounts to free energy contributions of

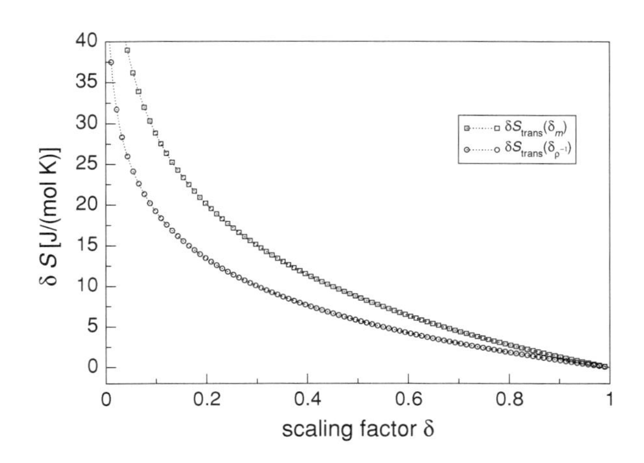

Fig. 3.7 Error of the translational entropy according to Eqs. 3.28 and 3.29 as a function of the corresponding scaling factor

$T\delta S_{\text{trans}}(\delta_m) = 8.6\,\text{kJ/mol}$ and $T\delta S_{\text{trans}}(\delta_{\rho_n^{-1}}) = 5.7\,\text{kJ/mol}$ for the scaling of mass and inverse number density, respectively. If for instance the transition from heavy water to normal water is exemplarily considered, a mass scaling factor of $\delta_m = 0.9$ is calculated. Thus, it is apparent from Fig. 3.7 that the error in translational entropy due to isotope effects or comparable mass inaccuracies can be neglected in most cases.

However, the situation could be different in solution. If non-covalent interactions of sufficient strength are present between solvent and solute, the effective translational mass could be considerably different from the molecular mass of the solute due to the translational motion of the whole solvent cage surrounding the solute. During the association reaction between two solute molecules, chemical transformations (e.g., the formation of new bonds, the alteration of functional groups, or the increase of the steric demand in the associated product) can lead to a mass reduction of the solvent cage per monomer, thereby introducing rather large uncertainties in the effective translational mass. Since these aspects are highly system-specific, Eq. 3.28 could help to estimate such effects for a concrete case. A similar situation is found for possible inaccuracies of the inverse number density ρ_n^{-1} as can be seen from Fig. 3.7. In a typical rrho computation, the number density of an ideal gas is applied in the frame of the Sackur–Tetrode equation. If this approach is to be employed for the treatment of condensed phase processes, the number density will be considerably different from that of an ideal gas, and the error in translational entropy can be estimated from Eq. 3.29 and Fig. 3.7, respectively. This can again be illustrated for the case of water at ambient conditions. An order-of-magnitude estimate of $\rho_n^{-1} = 1.8 \times 10^{-5}\ \text{m}^3/\text{mol}$ for the inverse number density of liquid water at room temperature can be obtained from the mass density ($\rho_m(H_2O) \approx 997\text{kg/m}^3$) and the molar mass ($M(H_2O) = 0.018$ kg/mol) [77]. In order to rescale the corresponding inverse number density of an ideal gas ($\rho_n^{-1} = 2.4 \times 10^{-2}\,\text{m}^3/\text{mol}$ at $T = 298.15$ K) to the value estimated for liquid water, a scaling factor of $\delta_{\rho_n^{-1}} = 7.5 \times 10^{-4}$ has to be applied, for which an error in translational entropy of $\delta S_{\text{trans}}(\delta_{\rho_n^{-1}}) = 59.8$ J/(mol K) is calculated according to Eq. 3.29. Such an error results in a considerable contribution to the free energy of about $T\delta S_{\text{trans}}(\delta_{\rho_n^{-1}}) = 17.8\,\text{kJ/mol}$ at $T = 298.15$ K and thereby constitutes a non-negligible impact on translational entropy changes in the condensed phase obtained from the rrho approach. It is important to note that according to Fig. 3.6 estimates of this kind are only reasonable at liquid phase densities for which translational length scales of approximately $a = 1 \times 10^{-10}$ m or larger are to be expected, because the error function expressed in Eq. 3.29 is derived from the Sackur–Tetrode equation (see Eq. 3.8), which in turn is based on the integral approximation for the evaluation of the translational partition function as expressed in Eq. 2.30. The analysis carried out in the previous part demonstrates that this approximation is of reasonable accuracy at translational lengths in the domain of about one hundred picometers and larger, and the order-of-magnitude estimate obtained in that section for liquid water at ambient conditions amounts to $a \approx 3 \times 10^{-10}$ m. These numbers indicate that the high temperature approximation and therefore the error function as expressed in Eq. 3.29 should be reliable for this particular case.

The second entropy contribution which will be examined in terms of the scaling factor approach is the rotational entropy expressed in Eq. 3.9. The evaluation of the error function δS_{rot} according to Eq. 3.27 for the special case of a symmetric top results in the following expression

$$\delta S_{\mathrm{rot}}(\delta_I) = \frac{3}{2} R \ln\left(\frac{1}{\delta_I}\right), \tag{3.30}$$

from which the corresponding expression for the general case of different principal moments of inertia can be obtained by ignoring the factor of three and treating each principal moment of inertia in terms of an individual scaling factor. However, for a general analysis of the error in rotational entropy the expression for a symmetric top in Eq. 3.30 can be employed without loss of generality. It is clear from Eqs. 3.30 and 3.28 that the error in rotational entropy has the same functional form as the error in translational entropy due to inaccuracies in the mass, which is based on the same functional dependancy of the rotational and translational partition functions on the principal moments of inertia and the mass, respectively (see Eqs. 2.30 and 2.36). However, the magnitude of the scaling factor δ_I will be different in general. Given that the mass of the investigated compound in general is available with a much higher accuracy, the inaccuracies in the principal moments of inertia obtained from quantum chemical calculations can be reduced to errors in the geometry, i.e., in the calculated bond lengths. Such errors largely depend on the applied quantum chemical methodology and the system under investigation. In the case of transition metal-ligand bonds, errors in the bond length of 15–20 pm have been reported for hybrid functionals commonly applied in quantum chemical calculations [17, 78, 79]. For structure optimizations of a single water molecule in terms of the Hartree–Fock method, a significant basis set dependancy of the oxygen–hydrogen bond length r_{OH} is observed [62]. While the result obtained from a double-ζ basis set ($r_{\mathrm{OH}} = 94.63$ pm) is quite close to the experimental bond length of $r_{\mathrm{OH}} = 95.78$ pm, the geometry obtained at the basis set limit yields an even smaller value of $r_{\mathrm{OH}} = 93.96$ pm [62, 80]. Corresponding values calculated from commonly employed density functionals yield values in the range of 96–98 pm in combination with double-ζ and triple-ζ basis sets [62]. From these examples a reasonable parameter interval of 0.8–1.2 for the rotational scaling factor δ_I can be estimated in order to calculate the error in rotational entropy due to inaccuracies in the applied quantum chemical methodology. A plot of the error function δS_{rot} in this interval is shown in Fig. 3.8.

From this graph it is apparent that structural errors typically arising in state-of-the-art quantum chemical applications do not exceed the magnitude of approximately 2.5 J/(mol K) (which is equal to a free energy contribution of 0.8 kJ/mol at $T = 298.15$ K) and thereby do not lead to considerable inaccuracies in the rotational entropy. However, the situation is again expected to be more complicated in the condensed phase. As in the case of the mass, the principal moments of inertia

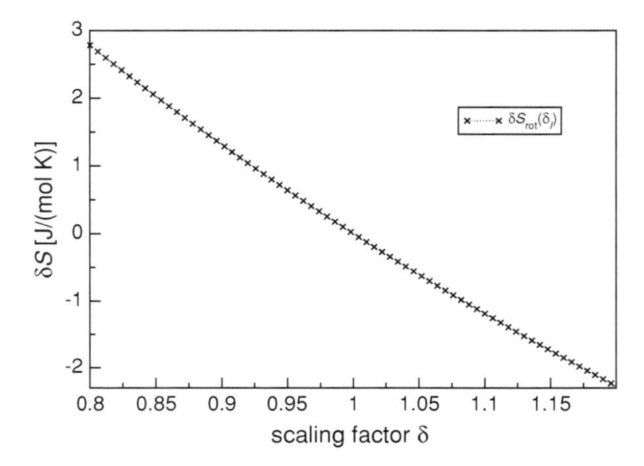

Fig. 3.8 Error of the rotational entropy for a symmetric top according to Eq. 3.30 as a function of the scaling factor δ_I

could be largely affected by non-covalent solute-solvent interactions and the resulting formation of a solvent cage. In addition, the bond distances of the solute will be influenced by such interactions, but also by the increased density of the surrounding medium, i.e., the values obtained from the isolated molecule calculation are expected to be subject to additional inaccuracies beyond the errors of the quantum chemical method. Thus, the effective moments of inertia of a compound in solution could require scaling factors from outside the interval considered in Fig. 3.8, but these influences are again difficult to quantify in a general manner and have to be estimated specifically for a given system e.g. in terms of microsolvation.

In contrast to the translational and rotational degrees of freedom, the scaling of the vibrational entropy contribution according to Eq. 3.27 results in an error function $\delta S_{\text{vib}}^{v_k}(\delta_{v_k})$ depending on the scaling factor δ_{v_k} as well as the scaled frequency v_k according to

$$\delta S_{\text{vib}}^{v_k}(\delta_{v_k}, v_k) = R\left[\frac{\beta h v_k}{\exp(\beta h v_k) - 1} - \frac{\beta h \delta_{v_k} v_k}{\exp(\beta h \delta_{v_k} v_k) - 1} + \ln\left(\frac{1 - \exp(-\beta h \delta_{v_k} v_k)}{1 - \exp(-\beta h v_k)}\right)\right]. \tag{3.31}$$

This dependancy on the scaling factor as well as on the frequency itself arises due to the exponential relation between frequency v_k and vibrational entropy S_{vib} as expressed in Eq. 3.10, which prevents a cancellation of the error-afflicted quantity in the error function as in the case of the logarithmic relation found for the translational and rotational entropies. The total error in vibrational entropy for a non-linear polyatomic molecule is then obtained as the sum of the individual contributions given by

$$\delta S_{\text{vib}}^{\text{tot}}(\{\delta_{v_k}\}, \{v_k\}) = \sum_{k=1}^{3M-6} \delta S_{\text{vib}}^{v_k}(\delta_{v_k}, v_k), \tag{3.32}$$

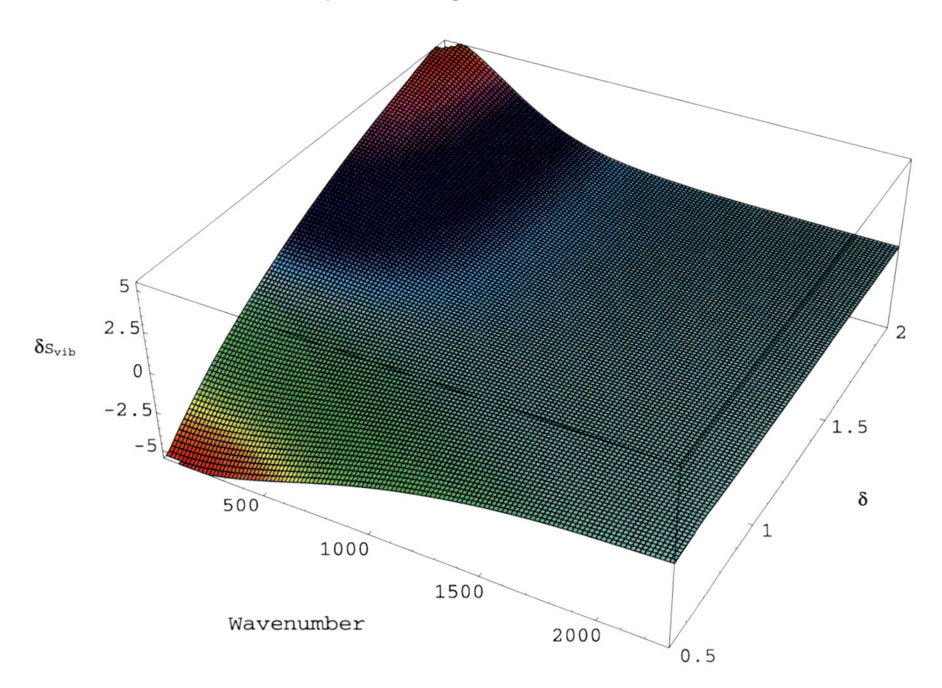

Fig. 3.9 Error of the vibrational entropy $\delta S_{\mathrm{vib}}^{\nu_k}(\delta_{\nu_k}, \nu_k)$ (in [J/(mol K)]) according to Eq. 3.31 as a function of the scaling factor δ_{ν_k} and the harmonic wavenumber $\tilde{\nu}_k$ (in [cm^{-1}]). The plot has been generated with the MATHEMATICA program [84]

where M again denotes the number of nuclei in the molecule. A comparison of different quantum chemical methodologies with respect to the performance in predicting harmonic frequencies shows that most quantum chemical methods routinely applied in standard calculations are able to reproduce the experimental harmonic frequencies of a single water molecule to values which correspond to scaling factors δ_{ν_k} between 0.9 and 1.1 [62, 81]. Similar values have been found in the case of more complex systems [82, 83]. Following these observations, the error in vibrational entropy $\delta S_{\mathrm{vib}}^{\nu_k}(\delta_{\nu_k}, \nu_k)$ is plotted in Fig. 3.9 as a function of the harmonic wavenumber and the scaling factor δ_{ν_k} in the interval between 0.5 and 2. This graphical representation indicates that in the chosen interval of the scaling factor δ_{ν_k} non-negligible errors in the vibrational entropy are only to be expected from individual modes of approximately $\tilde{\nu} = 500$ cm^{-1} or below. This corresponds to the previously obtained result that significant contributions to the vibrational entropy are only found for low-frequency vibrations, see the discussion in Sect. 3.2.1 and Table 3.12. Thus, for a reasonable error estimate of the vibrational entropy only those modes with wavenumbers of approximately $\tilde{\nu} = 500$ cm^{-1} and smaller have to be considered for the sum in Eq. 3.32. It is also clear that in this low-wavenumber domain values of the scaling factor δ_{ν_k} larger than 1 lead to an increase in vibrational entropy and vice versa. Therefore, a systematic scaling in terms of a single scaling factor $\delta_{\{\nu_k\}}$ normally done for the correction of inaccuracies in the applied quantum chemical method can lead to a considerable total error according to Eq. 3.32 if the number of low-frequency vibrations in the system under investigation is large enough. However, the impact of the inaccuracy of each individual mode on the free energy will typically not exceed ± 1 kJ/mol at $T = 298.15$ K if the scaling parameter lies within the interval estimated above to be relevant for the scaling of calculated harmonic frequencies. In contrast, even if individual scaling parameters $\{\delta_{\nu_k}\}$ of relatively large magnitude are applied to low frequency vibrations (i.e., the individual error in vibrational entropy $\delta S_{\mathrm{vib}}^{\nu_k}(\delta_{\nu_k}, \nu_k)$ of that particular vibration is of large absolute value according to Eq. 3.31), the overall effect on the vibrational entropy as expressed in the $\delta S_{\mathrm{vib}}^{\mathrm{tot}}(\{\delta_{\nu_k}\}, \{\nu_k\})$ value (Eq. 3.32) can be considerably smaller if these scaling factors in equal parts assume values larger and smaller than 1.

Such larger scaling factors can e.g. become important if anharmonic effects are to be modelled via this approach, but it should be noted that Eqs. 3.31 and 3.32 are only strictly valid in the case of harmonic vibrations, because they are based on the partition function of a harmonic oscillator. Furthermore, possible solvent-solute interactions in the liquid phase already discussed above for the translational and rotational entropy error functions might as well lead to vibrational scaling factors of larger magnitude as the upper (1.1) and lower (0.9) limiting values estimated for the correction of methodological deficiencies. As in the case of the principal moments of inertia, these interactions can result in larger effective masses involved in the harmonic oscillation, thereby reducing the actual frequency of the process and possibly leading to larger vibrational scaling factors as otherwise would be applicable for the corresponding vibration in the gas phase.

3.2.2.3 Technical Errors

Inaccuracies from the class of technical errors arise from the implementation of a certain electronic structure method in the form of a software code. The most important factors to be considered in this regard are the application of an incomplete basis set in the molecular orbital calculation (and the resulting contribution to inaccuracies in the calculated potential energy surface relevant for the computation of the principal moments of inertia and the normal mode analysis) as well as numerical errors due to the limited accuracy of hardware-based data processing or the application of numerical approximations to analytical expressions occurring in the applied methodology. An important example for the latter case is the calculation of harmonic frequencies as numerical derivatives of analytically obtained energy gradients as compared to the analytic evaluation of the (exact) second energy derivatives, which is a commonly applied procedure [61]. In most cases the impact of technical errors can be systematically reduced, e.g., by carrying out extrapolations to the complete basis set limit or by employing more sophisticated numerical differentiation procedures, but such approaches normally go along with an increase of the computational effort [61, 62, 85]. An example for such investigations is the examination of the basis set dependancy of the rrho zero point energy obtained from different density functional theory methods [86]. In this study it has been concluded (besides other things) that with the exception of a comparably small double-ζ basis set the effect of the chosen density functional is generally larger than the error introduced due to basis set incompleteness and that a basis set of triple-ζ size including polarization functions is normally sufficient even for the treatment of e.g. anharmonic effects.

With regard to the error analysis of the rrho approach, the effect of most technical errors can be handled in the same manner as the inaccuracies due to model-inherent errors treated in the previous part. The determination of appropriate scaling factors can be carried out in a straightforward way by performing model calculations with more sophisticated methods or basis set extrapolation studies and comparing to the original results. Accordingly, the framework for estimating the influence of technical inaccuracies on entropies from the rrho approach has already been derived in the last section and this class of errors will not be covered in further detail in this thesis.

3.2.3 Entropy Changes from the Rigid Rotor Harmonic Oscillator Model at Increased Densities

The final part of the present chapter will be focussed on a discussion of the applicability and accuracy of the rrho approach with regard to entropy changes in the condensed phase. The results and discussion presented in this chapter so far will provide a guideline for this rather widespread topic. In order to make this transition from the low density to the high density domain accessible in a comprehensible manner, the differences between a compound in the gas phase and in solution with regard to the different degrees of freedom will be reviewed shortly from a pragmatic and intuitive perspective, and a connection to the assumptions of the rrho model will be drawn. Afterwards, the main results of this chapter are summarized and a short discussion of the modeling of liquid phase entropy changes in the literature is presented, extending the literature review from Sect. 3.2.1.3 and introducing important concepts for a final evaluation of the rrho model at increased densities.

3.2.3.1 Transition from the Gas Phase to the Condensed Phase

In the condensed phase, it is reasonable to expect that completely free translations and rotations are no longer possible for an individual molecule. This is especially important if considerable solvent–solute interactions like e.g. hydrogen bonds are involved. Instead of the free motions, librations and hindered rotations can be expected to occur at these conditions. The quantum mechanical treatment of a hindered rotor is in principle possible but complicated, which is the reason why numerical tables of the energy eigenvalues as a function of the potential height V_0 between successive potential wells are normally applied in combination with the numerical computation of hindered rotor partition functions [21, 87]. The application of these approaches will not be investigated here for two reasons. First, the potential height V_0 is a highly system-specific quantity considerably depending on the solvent and the solute. There is no straightforward way to extract this information efficiently from static quantum chemical calculations, and extensive computations of rotation potentials via single point calculations in combination with a microsolvation treatment of the solvent soon become highly demanding in terms of the computational effort with increasing system size. An additional aspect of the system-specific nature of such a description for the translational and rotational degrees of freedom in solution is the temperature interval in which these degrees of freedom can reasonably be expected to behave like hindered rotations. At low temperatures where $k_BT \ll V_0$ is fulfilled, most of the molecules will be concentrated at the potential wells and the crossing of the potential barrier V_0 can be expected to be a rare event. Thus, the effective potential for the rotational movement is accurately described by a parabolic form and the corresponding degree of freedom is accurately represented by a harmonic oscillation, making the additional difficulties of the treatment as a hindered rotor essentially unnecessary. In the case of high temperatures or low potential heights (i.e., $k_BT \gg V_0$), most molecules are not significantly constrained by the presence of the changes in the rotational potential, and for that reason the corresponding degree of freedom could be approximated as a free rotation without considerable loss of accuracy. According to these considerations, the combination of a model which includes hindered rotations with the static quantum chemical approach would lead to a more complicated formalism and introduce a considerable degree of system specificity. The second reason for disregarding a more detailed treatment in terms of hindered rotations is related to the purpose of the present chapter, which is an assessment of the rrho approach at high densities, i.e., a validation of the approximations introduced by the rrho model at conditions it has not been developed for originally. The consideration of major extensions like the application of a different underlying quantum mechanical model system would alter the rrho approach considerably and therefore abandon the basic ingredients of the model, thereby making a concise assessment of the original approach nearly impossible. With regard to the results presented so far, a scaling of the rigid rotor entropy in terms of increased effective principal moments of inertia via the scaling approach introduced in the previous section (see Eq. 3.30) would be more helpful.

Another issue apparently relevant for the transition from the gas to the condensed phase is the increased importance of intermolecular interactions with regard to the factorization of the N-particle partition function. This aspect is of special relevance if relatively strong intermolecular interactions like hydrogen bonds are present in the system under investigation, but will be non-negligible in most other cases as well. One suitable approach for the treatment of such interparticle effects lies in the explicit treatment of an interparticle potential function and the evaluation of expressions derived from the (classical) N-particle partition function (see Eq. 2.5) as outlined in Sect. 2.3. However, this course of action is again not combinable with the fundamental idea of the partition function factorization found in the rrho approach (see Eq. 2.26), which per definition is only exact in the absence of an interparticle interaction potential. Nevertheless, a straightforward extension of the rrho model concerning the treatment of interparticle interactions is given by the quantum cluster equilibrium (qce) model outlined in Sect. 2.2.1. In this approach, the interactions between different particles to a large extent are treated via the quantum chemical calculation (i.e., clusters are the fundamental units instead of the single particle), thereby increasing the computational effort in a considerable but for many systems still feasible way. These clusters are treated as a non-interacting gas in the zeroth order qce model, which permits the usual factorization of the N-particle partition function into contributions from the different cluster degrees of freedom closely resembling the corresponding expressions of the conventional rrho approach, see Sect. 2.2.1. In a refined treatment residual interactions between different clusters are modelled in terms of a mean-field approach, thereby affecting the electronic partition function and some other model-specific expressions, but not preventing the factorization of the N-particle partition function. Thus, the qce model can be understood as an extension to the rrho model for the treatment of interparticle interactions, which leaves the basic assumptions and approximations of the original approach untouched. The performance of the qce model for the calculation of thermodynamic quantities and especially entropies in the condensed phase will be investigated in detail in the following chapters of this thesis.

The third contribution to which an effect due to the transition from the gas to the condensed phase might intuitively be expected are the vibrational degrees of freedom. However, as already indicated in Sect. 3.2.2, the overall impact of a possible frequency shift between gas phase and solution might be rather small as compared to other factors, but since these shifts can be determined experimentally a direct quantification is possible [88–90]. A summary of several frequency shifts for the transition from the gas to the condensed phase measured for a number of small molecules and different solvents can be found in [89]. These numbers indicate that shifts of this kind typically assume magnitudes in the range between 5 and 30 cm^{-1}, and that the sign of these shifts is not necessarily uniform for the set of vibrational modes in a given molecule. Largest shifts of -43 and -44 cm^{-1} are found for the hydrogen–oxygen stretching vibrations of water in a chloroform solution. The translation of these shifts into individual vibrational scaling factors δ_{ν_k} as introduced in the frame of the error analysis presented in Sect. 3.2.2 results in values of $\delta_{\nu_k} = 0.99$, from which a negligible effect on the vibrational entropy is predicted according to the plot in Fig. 3.9. Even in the case of low-frequency vibrations of about 550 cm^{-1} (methyl iodide dissolved in itself), shifts of approximately 20 cm^{-1} correspond to values of vibrational scaling factors as low as $\delta_{\nu_k} = 1.04$, thereby again affecting the vibrational entropy to virtually no extent [91]. Thus, as long as the number of shifted low-frequency modes is not too large and the sign of these shifts is not uniform, the effect on the vibrational entropy due to the transition from the gas to the liquid phase is considerably lower than the those of the remaining degrees of freedom and can be neglected as a first approximation.

The final and putatively most important quantity considered for the transition to the liquid and the application of the standard rrho model in the condensed phase is the volume relevant for the translational degrees of freedom. Besides the possible occurrence of hindered rotations for a molecule in solution, the translational volume has been identified before in this chapter as a crucial quantity for the transition from the gas to the condensed phase. This is evident, because the major difference between these two phases is the difference in density, which directly affects the volume of the system if the particle number is kept constant. Furthermore, the computational results obtained for the association reactions in the present chapter demonstrate that, regardless of the size of the investigated system, the translational degrees of freedom always yield the largest contributions to the total entropy change, see Tables 3.11 and 3.7. These observations are in total agreement with the numerous publications treating the entropy change of association reactions in solution, some of which will be discussed in greater detail in the following part [34, 36, 39, 41, 42, 92–97]. As pointed out in Sect. 2.1.2, the basis of the volume dependancy of the translational entropy contribution according to Eq. 3.8 is introduced via the treatment of the translational degrees of freedom in the frame of the particle in a cubic box model. In this approach, the size of the box $V = a^3$ defines the portion of space in which the particle of mass m can undergo a potential-free translational motion. Outside this area the potential is assumed to be infinitely high, and the probability of finding the particle outside the box is therefore zero. The analogy to a particle in a dilute gas confined to a box of some macroscopic dimension is striking, but the question arises how this picture changes upon the transition to the liquid state. The first and probably most straightforward conclusion would lie in the estimation (or calculation) of an effective box length, in which translational motion in the condensed phase is actually taking place. Such estimates have been proposed previously (see Sect. 3.2.2) to assume orders of magnitude of several hundreds of picometers in the case of liquid water at ambient conditions. Even if such estimates only constitute a rough appraisal and possibly rely on empirical data as e.g. the density of the liquid phase under consideration, their incorporation into the rrho approach can be done in a straightforward way for instance via Eqs. 3.15 and 3.29 in contrast to more elaborate refinements like the hindered rotor model, which in addition violate the basic assumptions of the rrho approach. Corrections of this kind to the translational entropy in solution have been termed free volume theories of translational entropy recently and are usually based on a cell model of liquids [96–100]. However, there are also arguments in literature which propose that such a partitioning of the system volume into individual contributions for each molecule is not reasonable, at least if a comparison with experimental measurements is intended [97]. According to these objections the *instantaneous* translational volume available to a molecule in solution is indeed given by the dimension of the solvent cage, but over the time scale of typical experimental measurements the molecule will still undergo translational motion within a volume comparable to the macroscopic dimensions of the containing vessel. Thus, the translational volume which has to be employed for the evaluation of the Sackur–Tetrode equation should still be that of the whole system. If this reasoning is applied consistently, an explicit time dependancy of the translational volume ought to be introduced according to $V \rightarrow V(t)$, and the actual value of $V(t)$ has to be obtained from system-specific properties like e.g. the diffusion coefficient. However, the question arises if such a treatment is still consistent with the original model, namely the particle in a cubic box. The quantum mechanical treatment employed for the calculation of the energy eigenvalues in Eq. 2.28 is based on the stationary Schrödinger equation, and the box length a is usually considered to be constant [22]. In this case, the Hamilton operator does not explicitly depend on time, and the separation of the time-dependent Schrödinger equation into the time-independent equation and a time-dependent phase factor to the stationary eigenfunctions can be carried out. Thus, the application of the energy eigenvalues expressed in Eq. 2.28 is based on a stationary treatment of the problem, and the length scale available for translational motion is thereby best understood as an instantaneous property of the system not depending on time.

According to these considerations, a modified treatment of the translational degrees of freedom in terms of a corrected number density ρ_n is most important for the adaption of the rrho model to entropy changes in the condensed phase. The consideration of hindered rotations, increased interparticle interactions, or shifts in vibrational modes either results in a significant complication of the formalism or leads to entropy changes which are expected to be negligible is most cases. This outstanding influence of the translational degrees of freedom has been recognized in the literature before, and the following part will provide a short overview of studies related to this subject.

3.2.3.2 Approaches for Modeling Condensed Phase Translational Entropy Changes in the Literature

As already pointed out in the previous part as well as in Sect. 3.2.1, the transfer of gas phase entropies to systems in solution and the special importance of the translational degrees of freedom is a widely discussed subject in the literature, especially in the field of biochemistry [34, 36, 39, 41, 42, 92–97, 101]. Several approaches have been developed and proposed in the last decades, most of which are covered in the review articles found in [92, 97]. In the following, several of these methods as well as some of those already introduced in Sect. 3.2.1 will be discussed shortly with regard to a possible applicability in the frame of the rrho approach.

One of the earliest approaches to the problem was given by Steinberg and Scheraga, who concluded that translational entropies according to the Sackur–Tetrode equation are also applicable to condensed phase processes [34]. This suggestion is based on the assumption that translational and rotational motions in solution are equal to the corresponding motion in the gas phase and that the effect of the solvent on the entropy of association can be considered as a separate term independent from the contribution of translation and rotation. Such a treatment considers the solvent as a structureless continuum not affecting the translational and rotational degrees of freedom, and the solvent contribution to the association entropy is defined as the difference between the experimental value and the *intrinsic* entropy change, which is equal to the result obtained for the corresponding gas phase reaction. It is apparent that this approach is not capable of providing the actual entropy contribution arising from the solvent-solute interaction without experimental aid (though methods for estimating this part are given for the case of protein association reactions) and therefore is not well suited for computational applications of the rrho model. Nonetheless, the method and several variations have been applied in various studies [102, 103].

Another approach already discussed in Sect. 3.2.1 equates the effect of the solvent on the entropy of a solute to the vaporization entropy of that substance [36, 38]. This correction can be understood as a two-step process in which the solute is transferred from the gas phase to the pure liquid state and afterwards diluted to the actual concentration in solution. For the association reaction between two urea molecules in liquid water it is found that the association entropy obtained from this model is reduced by approximately 50% as compared to the ideal gas value, i.e., upon the transfer from the gas phase to a 1 M aqueous solution [38]. Such correction schemes based on an empirical treatment of the condensation process are in principle well suited for a combination with the computational rrho approach, which for instance has been shown for the case of the Wertz-model (discussed in Sect. 3.2.1) by Zhu and Ziegler [56, 59]. However, the application of these models often requires knowledge about reference quantities (like the vaporization entropy of the solute in the approach proposed by Doig and Williams), which in general are not predictable by the methods of first principles quantum chemistry. In the case of the vaporization entropy, a reasonable guess can be expected from Trouton's rule if the liquid phase of the solute under investigation is not associated significantly, but the general applicability of these empirical corrections might be severely limited with regard to chemically complex systems [104].

Besides the improvement of the model description on the side of the solute in the liquid phase, some effort has also been raised to obtain more accurate predictions for entropy changes in solution via a consideration of contributions to such changes in terms of the solvent. An early suggestion of this kind was proposed by Gurney and has become known as the cratic correction term, which is given by [105, 106]

$$S_{\text{cratic}} = \Delta n R \ln\left(\frac{1{,}000\rho_s}{M_s}\right), \tag{3.33}$$

where ρ_s and M_s denote the density and molar mass of the solvent, respectively, and Δn the change in particle number during the reaction course. The cratic correction can be understood as the solvent contribution to the entropy of the solute arising from the solvation process, i.e., it is a measure for the entropy due to the mixing between solute and solvent at a standard solute concentration of 1 M in the case of an ideal solution. For a particle conserving reaction (i.e., $\Delta n = 0$), there is no need for such an approach, since the mixing entropy according to Eq. 3.33 solely depends on solvent properties which are not involved in the reaction. However, for an association (or dissociation) reaction where $\Delta n \neq 0$, the cratic correction has been used to "subtract out" the additional translational entropy contributions appearing on the reactant's side of the reaction [106, 107]. It has been argued that this additional translational motion is reasonably reflected in the additional mixing entropy arising from the increasing (or decreasing) particle number, and that the total entropy balance corrected via this approach refers to a system in which no effective center of mass translational motion takes place, i.e., to an entropy change for translationally stationary particles. Such estimates for translational association entropies in solution could be brought to a better agreement with experimental measurements, and it was proposed to prefer the cratic entropy contribution over corresponding values obtained from the Sackur–Tetrode equation and over the vaporization entropy correction as well [107]. However, a detailed analysis of Holtzer demonstrates that there is no rigorous basis for the cratic correction either in thermodynamics or in statistical mechanics, and that the situation of stationary reaction participants putatively accessible via Eq. 3.33 is not fulfilled in this approach [39]. In addition, a warning concerning the artificial partitioning of entropy into contributions from center of mass translation and rotation as well as internal vibrational degrees of freedom is given, and it is pointed out that in a proper statistical mechanical treatment all momentum contributions to the free energy change (and thereby also to the change in entropy) necessarily have to cancel. This is rationalized in terms of a viewpoint according to which a molecule is understood as a sum of atoms each possesing three translational degrees of freedom, and that the number of atoms is constant during each chemical reaction, no matter if the number of molecules does change [39]. It is clear that such a sum-over-atoms perspective conveniently replies to the methodology of md simulations, but is not compatible in a straightforward way to the assumptions of the rrho model, which explicitly relies on a partitioning into different degrees of freedom.

From Eqs. 3.13 and 3.15 it is apparent that this momentum (mass) independence is not given for entropy changes in the frame of the rrho model at first glance. However, this point has been examined in detail as a part of a recent study, and it could be demonstrated that even if each individual contribution from the rrho model to the entropy change depends on the mass, this dependancy almost cancels in the total entropy change in the case of the quantum treatment of vibrations and exactly cancels for a classical treatment of the vibrations, see Fig. 3 and Sect. 5.3.2 in [97]. In the case of the sum-over-atoms (flexible molecule approach, fm), there is no mass dependancy as expected according to the reasoning pointed out by Holtzer [39, 108]. Thus, the decomposition into molecular degrees of freedom can be understood as being artificial in the sense that the physical meaning of each kind of degree of freedom can be questioned due to an apparent dependancy on momentum quantities, but the total change in entropy is consistent with regard to this objection. Nevertheless, the fundamental problem of the large sensitivity of reaction entropies from the rrho model to the stoichiometry of the reaction is of course also present in the total entropy change and therefore not subject to these aspects.

The final suggestion from the literature to the problem of predicting (translational) entropy changes in solution discussed here with regard to a possible application in terms of the rrho model will be the free volume theory of translational entropy already referred to previously [96, 97, 101]. The basic idea behind this approach is related to the cell theory of liquids, in which the total liquid phase volume V is divided into a large number of equal cells each of which contains a single molecule [97, 98, 100]. The individual cell volume v is given in a straightforward way as $v = V/N = \rho_n^{-1}$, i.e., it is equal to the inverse number density already introduced in Sect. 3.2.1. In contrast, the free volume v_f for an individual particle is given in terms of a localized configuration integral (see Eq. 2.5) according to [96]

$$v_f = \int\limits_V \exp(-\beta U(\mathbf{r}))d\mathbf{r}', \tag{3.34}$$

where $\int_V d\mathbf{r}'$ denotes an integration over the whole cell volume v. As outlined in Sect. 2.1.1, this expression reduces to the cell volume v in the limiting case of non-interacting particles of zero volume and translational motion in a potential-free box (cell), and in cases where the evaluation of Eq. 3.34 is not possible due to missing knowledge of the interaction potential $U(\mathbf{r})$ an estimate of the free volume v_f in terms of the cell volume v and the spherical molecular diameter is provided in [96]. Following this reasoning, the total configurational integral Z is given as $Z = (Nv_f)^N$, and the translational entropy of a particle in the liquid phase is obtained from a Sackur–Tetrode-like equation [96]

$$S_{\text{trans}}^{\text{liq}} = Nk_{\text{B}} \ln\left(\frac{v_f}{\Lambda^3}\right) + \frac{5}{2}Nk_{\text{B}}, \tag{3.35}$$

i.e., the particle in solution is treated like a gas with an effective smaller free volume exponentially weighted in terms of the interaction potential $U(\mathbf{r})$. From a principal point of view a combination between the treatment of translational entropy in the rrho approach and this free volume model can be accomplished in a straightforward way. However, in the case of the association reaction between a protein and a ligand (for which the model was originally developed for) a different equality for the translational entropy change as the one proposed in Eq. 3.12 has been employed, which is given by [96]

$$\Delta S_{\text{trans}} = S_{\text{trans}}^{\text{protein,b}} + S_{\text{trans}}^{\text{ligand,b}} - S_{\text{trans}}^{\text{protein,f}} - S_{\text{trans}}^{\text{ligand,f}}. \tag{3.36}$$

In this equation the superscripts b and f indicate the bound and the free state of the protein and the ligand, respectively, i.e., the associated complex is still treated as two individual particles, but the translational volume of the ligand is considerably reduced upon association, namely to the (estimated) volume of the binding site. A similar approach has been suggested independently of the cell theory model in [97]. The basic difference to the conventional rrho model lies in the fact that there is no interconversion between different degrees of freedom according to Eq. 3.36, because the ligand bound to the binding site still possesses three translational degrees of freedom and no new vibrations are introduced in the complex. This is rationalized in terms of the non-covalent binding situation of the ligand, which more closely represents a translational motion according to the particle in a box model than a harmonic vibration [96]. In addition, the estimation of the frequencies required for a harmonic oscillator treatment has been deemed to be difficult [96]. However, the situation might be different in many chemical applications, where the investigated structures are usually smaller and all harmonic frequencies are accessible via a normal mode analysis. Furthermore, if a covalent bond is formed upon association, a harmonic oscillator treatment is probably more precise than the particle in a (potential-free) box approach. It should also be noted that the form of Eq. 3.36 is similar to the exchange formation approach introduced for the treatment of the pseudorotaxane association reaction in solution, see Eq. 3.2. In combination with the additional consideration of a different particle density of solute and solvent in terms of the free volume v_f and of an empirical estimate of the rotational entropy change, this free volume approach has been successfully employed for the prediction of the liquid phase entropy of pure water, and rate enhancement values for intramolecular binding processes as compared to the corresponding intermolecular reactions have been reproduced to the right order of magnitude [96]. In addition, the necessity of a comparable correction for the rotational entropy change has been stressed.

Thus, translational entropy changes of association reactions in solution are a wide-spread problem occurring in many fields of the chemical sciences, and several approaches have been suggested for its theoretical treatment. Some of these have been reviewed in this part, with a special emphasis on those methods probably applicable for a combination with the rrho model and computational first principle approaches. Consequently, the following part will take a closer look at the practicability of such a combination for the calculation of translational entropy changes in solution, which ideally does not depend on the stoichiometry of the underlying reaction.

3.2.3.3 A Final Assessment of the Rigid Rotor Harmonic Oscillator Model for Entropy Changes in the Condensed Phase

The discussion in this chapter and the presented results demonstrate that a theoretical treatment of entropy changes of association reactions in solution is a demanding and complex procedure, no matter what (or if any) computational methodology is applied for the generation of molecular data. The two parts reviewing the suggestions made in previous studies on this subject (see Sects. 3.2.1.3 and 3.2.3.2) indicate that even the interpretation of experimental data or the specification of most essential quantities and concepts are controversial issues. The rrho model as outlined in Sect. 2.1.2 certainly is one of the most simple approaches to this problem and it might even be questioned if this high temperature gas phase approximation to statistical thermodynamics is suitable for the treatment of such complex processes at all. However, from the perspective of first principles quantum chemistry this approach is valuable particularly due to its simplicity and the straightforward combination with electronic structure calculations. As pointed out before, such computations yield an important contribution to fields of research as e.g. homogeneous catalysis, where an accurate treatment of the electronic structure might be an important issue due to the chemical complexity of the involved species and force field-based methods for the calculation of thermodynamic quantities (as discussed in Sect. 2.3) are not applicable. A model for translational entropy changes in solution based on the simplicity of the rrho approach would therefore be very useful. It is apparent that due to the complexity of the association process in solution a treatment via the quantum chemical single molecule approach has to rely on an empirical extension of the rrho model aligned to the practical applicability in standard calculations. All approaches from the literature discussed in the previous part can in principle be combined with the rrho model, but a correction in terms of vaporization entropies might be less useful in the case of chemically complex (or novel) compounds for which this quantity might not be known. In contrast, the alternative modification of the translational entropy according to the cratic correction term is explicitly formulated for a 1 M solution (even though a generalization to arbitrary concentrations would be straightforward) and thereby introduces a quantity which at first is not given in the single molecule picture of static quantum chemical calculations. It is also clear that the application of the cratic correction is limited to the association process in solution and would not be reasonable for the corresponding process in the gas phase, thereby leading to a formal discrimination between the two density domains. More important, it has been shown that there is no sound basis for this approach from a theoretical perspective, which are the reasons why the cratic correction will not be considered in this thesis [39]. Thus, a correction of the translational entropy change in terms of the free volume approach will be carried out in the following and evaluated on the basis of the experimental entropy change measured for the association of the Vögtle-type pseudorotaxane **V2** in chloroform solution ($T\Delta S = -8.8$ kJ/mol at $T = 303$ K, see Sect. 3.1.3) [1, 96, 101]. It is apparent from Eq. 3.35 that the translational entropy in this approach is very similar to the conventional Sackur–Tetrode equation employed in the rrho model. However, it should be noted that this approach has not been derived in the frame of the rrho model, in which a true interconversion of degrees of freedom takes place and the translation and rotation of one of the monomers are treated as harmonic vibrations in the associated complex. Rather, the free volume approach has been evolved from the theoretical treatment of non-covalent protein-ligand associations in solution, for which the association entropy is calculated according to Eq. 3.36 and the residual center-of-mass motion of the ligand is still treated as translational and rotational degrees of freedom [96, 97]. According to that, the translational entropy change does not depend on any momentum quantity and is *not* given by an equation of the form as in Eq. 3.15, which clearly depends on the mass relation of the reacting species as expressed in Eq. 3.16. The free volume approach has been combined with the rrho model only to account for the reduced size of the active site in a metallo-enzyme or to give qualitative reasons for large entropy losses of association reactions in solution, but not for the consideration of effective translational length scales of a solute in a specific solvent in combination with a translational entropy change according to Eq. 3.15 [109–111]. Thus, it will be interesting to see if the application of the free volume model within the conventional rrho approach is able to affect the calculated association entropies significantly and if a result as accurate as in the case of the exchange formation of the pseudorotaxane association (see Table 3.10) can be obtained, which explicitly circumvents the interconversion of different degrees of freedom.

According to the discussion in the previous section, it is clear that the basic difficulty for the application of the free volume approach is the evaluation of the free volume v_f given in Eq. 3.34. Even though the spatially restricted integration constitutes a major simplification over the exact equality (see Eq. 2.5), there is no straightforward way to sample the potential of a particle within its cell via static quantum chemical methods. For this reason the determination of v_f has to be accomplished in an empirical way, and one possible solution to this problem has been suggested in [96]. In this approach the particle in solution is treated as a hard sphere with diameter d, and the free volume v_f in solution can be estimated from the volume per particle v obtained from the number density ρ_n of the liquid and the the diameter d according to

$$v_f = (v^{1/3} - d)^3. \tag{3.37}$$

A similar method (ignoring the volume of the particle introduced via d) has been employed in Sect. 3.2.2 for estimating the translational length scale in liquid water to a value of approximately $a = 3 \times 10^{-10}$ m. This will be an upper bound to the true translational length in water due to the additional effect from the volume of the particles not considered in this estimate, and a more refined value obtained from traditional md simulations ($a = 4.6 \times 10^{-11}$ m) is nearly one order of magnitude smaller than this estimate [101].

Following this approach, the estimated free translational length scale in the case of liquid chloroform is equal to $a = 1.5 \times 10^{-10}$ m based on a number density of $\rho_n = 2.4$ mol/L at $T = 298.15$ K and an estimated diameter of $d = 3.6 \times 10^{-10}$ m calculated as twice the length of a Cl–C bond in the chloroform molecule (the bond length is obtained from a geometry optimization employing the same quantum chemical methodology as in the case of the pseudorotaxane calculations, see Sect. 7.1) [112]. However, with regard to the determination of v_f for the pseudorotaxane components it is important to note that the free volume of a particular solute is mainly influenced by the interaction with its neighbor molecules and not by its size, because v_f is a measure for the restricted motion of the *center of mass* of the solute and not for the whole molecular structure [101]. This is clearly apparent from the v_f values obtained from md simulations in [101], according to which e.g. the computed average free volume for glycerol in water is smaller than that of a single water molecule in water. In addition, it is pointed out that the translational motion of a large solute requires the coordinated movement of a larger number of solvent molecules, which is the reason why spatially extended molecules tend to exhibit *smaller* effective v_f values as opposed to possible expectations based on the size of the solvent cage alone [101]. Thus, the translational length scale of the macrocycle **V** (or one of the axles **1–5**) will probably be even smaller as the value estimated for a single chloroform molecule. The effect of the solvent-solute interactions on the translational length scale can be approximated by comparing the intermolecular hydrogen-chlorine distance r between a pair of chloroform molecules ($r = 4.8 \times 10^{-10}$ m) to the distance between the chloroform hydrogen atom and the oxygen atom of the amide group coordinating the solvent molecule in complex **V6** ($r = 2.0 \times 10^{-10}$ m, see Table 3.1). From these values a distance scaling factor of approximately 0.42 can be estimated, and according to that a reduced translational length scale of $a = 6.3 \times 10^{-11}$ m for the Vögtle-type pseudorotaxane components in chloroform solution.[8] If this value is taken as an estimated reference for the pseudorotaxane association in chloroform solution, the effect on the entropy according to the direct formation (see Eqs. 3.1 and 3.12) can be quantified directly in terms of the scaling factor approach introduced in Sect. 3.2.2.2. The free volume of a particle in an ideal gas applied in the conventional Sackur–Tetrode equation (see Eq. 3.8) is given by $v_f = V/N_A = 4.1 \times 10^{-26}$ m^3, i.e., it is equal to the inverse number density ρ_n^{-1} per particle. From the estimated translational length in chloroform solution ($a = 6.3 \times 10^{-11}$ m) a free volume of $v_f = 2.5 \times 10^{-31}$ m^3 for the pseudorotaxane components is calculated, which corresponds to a scaling factor of $\delta_{\rho_n^{-1}} = 6.2 \times 10^{-6}$ for the scaling of the inverse ideal gas number density. This scaling factor can directly be employed for the calculation of the translational entropy error with the aid of Eq. 3.29, thereby yielding a difference of $T\delta S_{\text{trans}}(\delta_{\rho_n^{-1}}) = +29.7$ kJ/mol at $T = 298.15$ K. Thus, the magnitude of the translational entropy loss upon association is considerably reduced by this free volume estimate. Table 3.15 summarizes the thermodynamic data of the direct formation model (see Eq. 3.1) calculated according to the free volume approach for the translational entropy change.

Besides the model-inherent volume scaling to the estimated v_f-value in chloroform solution, the additional influence of the high temperature approximation in

[8] Please note that this is only an estimate relying on the assumption that the distance of the solvent molecules to all parts of the solute is equal to the corresponding distance calculated for the axle-to-wheel hydrogen bond in the solvent complex **V6**.

Table 3.15 Calculated thermochemical quantities for the direct formation reaction at $T = 298.15$ K and $p = 101{,}325$ Pa as well as the experimental free energy change of association ΔG_{exp} measured at $T = 303$ K [1]

No.	Guest	Direct formation			Experiment
		ΔH	$T\Delta S_{v_f}$	ΔG_{v_f}	ΔG_{exp}
V1	R-NO_2	−26.9	−21.7	−5.2	−13.7
V2	R-Cl	−28.1	−20.2	−7.9	−13.6
V3	R-H	−28.2	−20.0	−8.2	−11.0
V4	R-^{t}Bu	−28.2	−19.1	−9.1	−11.4
V5	R-OCH_3	−29.7	−18.6	−11.1	−12.1

All entropy contributions $T\Delta S_{v_f}$ are corrected for the estimated free volume effect ($T\delta S_{\text{trans}}(\delta_{\rho_n^{-1}}) = +29.7$ kJ/mol) and the effect of the high temperature limit $T\delta S_{\text{trans}}^{\text{num}} = +0.7$ kJ/mol. All values in [kJ/mol]

the derivation of the Sackur–Tetrode equation (as well as in the derivation of Eq. 3.35) has to be considered.

From the data set plotted in Fig. 3.6 a relatively small shift of approximately $T\delta S_{\text{trans}}^{\text{num}} = 0.7$ kJ/mol at $T = 298.15$ K is obtained for the transition from the integral approximation to the numerical summation result at a translational scale of $a = 6.3 \times 10^{-11}$ m, which nonetheless is considered for the values listed in Table 3.15. These corrected results indicate that the treatment of the translational entropy contributions in terms of the free volume approach instead of the conventional Sackur–Tetrode equation change the predicted overall thermodynamic situation completely. The entropy penalty at $T = 298.15$ K upon association is now smaller in magnitude than the corresponding enthalpy changes for all substituents, and therefore the association reaction to the pseudorotaxane complexes is predicted to be exergonic in all cases. This qualitative prediction is in agreement with the experimental measurements and also is in line with the results obtained from the particle conserving exchange formation reaction, see Table 3.10. From a quantitative perspective it is apparent that the calculated ΔG_{v_f} values more accurately reproduce the experimental values and even lie within the experimental error bars (approximately ± 2 kJ/mol) in some cases, whereas the numbers obtained from the exchange formation model overestimate the complex stability to a larger extent on average. However, with regard to the van't Hoff analysis of the association between axle **2** and the macrocycle ($\Delta H_{\text{exp}} = -22.0$ kJ/mol, $T\Delta S = -8.8$ kJ/mol at $T = 303$ K) the association entropies predicted by the exchange formation can be classified as being more accurate and the values listed in Table 3.15 obviously benefit from a cancellation of errors, in which the magnitudes of both the association enthalpy and entropy are overestimated but compensate each other partly to yield a free energy change close to the experimental results. Thus, the free volume model of the translational entropy in solution provides a significant improvement over the unmodified Sackur–Tetrode prediction, but the model of the particle conserving exchange formation apparently is even closer to the real situation in the case of the entropy change. However, this performance is validated only for a single measurement, and it is not clear how important the combined effect of the neglected contributions arising from e.g. hindered rotations or interparticle interactions as well as the model-inherent shortcomings due to the applied quantum chemical methodology really are for the transition from the gas to the liquid phase. For instance, the consideration of the different free volumes v_f estimated for the solutes ($v_f = 2.5 \times 10^{-31}$ m^3) as well as for a single chloroform molecule ($v_f = 3.4 \times 10^{-30}$ m^3) gives rise to a small but non-negligible entropy contribution of $T\delta S_{\mathrm{trans}}(\delta_{\rho_n^{-1}}) = +6.5\,\mathrm{kJ/mol}$ at $T = 298.15$ K for the exchange formation as well, which takes the values listed in Table 3.10 close to ΔS-neutrality and results in an additional stabilization of the associated products. If this contribution is considered, the free energy changes according to the exchange formation are no longer as accurate as the values from Table 3.15, which indicates that either the difference in free volume of solute and solvent molecules is smaller than the estimated value or that the combined magnitude of the other contributions (see Sect. 3.2.3.1) affecting the free energy change is larger than expected. In order to arrive at a final assessment of the rrho approximation with regard to the calculation of entropy changes in the condensed phase, a summary of the important results and conclusions obtained in this chapter so far will be given in the following.

- The comparison between experimental association entropies measured in the gas phase and those obtained from the rrho model as expressed in Eq. 3.12 shows that the theoretical predictions are accurate (see Table 3.11) even though an interconversion of different degrees of freedom is observed according to the stoichiometry of the association reaction. This indicates that the considerable loss of entropy upon association is correctly reproduced by the model, but due to the artificial partitioning into (momentum-dependent) contributions this is not necessarily based on a translational motion of the center of mass as predicted by the Sackur–Tetrode equation (see Eq. 3.8).
- The entropy changes computed from the unmodified rrho model for the association of the pseudorotaxane complexes are in reasonable agreement to the experimental values measured in solution as well if the number of particles is artificially kept constant during the association process in terms of microsolvation (see Table 3.10). Thus, the rrho approach is in principle capable of reproducing the entropy change in solution if no interconversion of degrees of freedom takes place.
- The formation of new low-frequency vibrations in the associated compound partly compensates for the unfavorable rotational and translational contributions, but the magnitude of this compensation is to small to get the overall process close to ΔS-neutrality. Within the rrho model only low-frequency modes yield relevant contributions to the entropy, but in all examined cases the newly formed vibrations can all be classified as low-frequency modes (see Table 3.12).
- The large magnitude of the translational entropy change is the result of even larger absolute contributions arising from a mass dependent term and a number density dependent term in the equality obtained for the translational entropy change according to Eq. 3.12. Due to the difference in sign a smaller value

results for the total translational contribution, which nonetheless is always larger than the values calculated for the rotational and vibrational degrees of freedom (see Table 3.13).

- The approximations and errors occurring in the combined application of the rrho model and first principles quantum chemical calculations can be assorted in different error classes. Whereas the effect of model errors is difficult to estimate in most cases, a quantification of model-inherent errors in terms of scaling factors and entropy error functions as expressed in Eq. 3.27 is readily obtained. These scaling factors may either be used to calculate (and correct) possible shortcomings of the applied methodology or to model a more drastic change of the particular quantity arising from external influences (like e.g. the transition from the gas to the liquid phase).
- A quantification of model errors has only been done for the approximations occurring in the derivation of the Sackur–Tetrode equation, see Sect. 3.2.2.1. It is observed that at almost all conditions relevant for a molecular system these approximations can be classified as being accurate. The integral approximation applied in the high temperature limit for the evaluation of the translational partition function is accurate down to translational length scales of $a = 1 \times 10^{-10}$ m, but in the limit $a \rightarrow 0$ predicts physically unreasonable translational entropies (see Fig. 3.6).
- From an intuitive point of view effects due to the transition from the gas to the liquid phase might be expected for the rotational motion, the degree of interaction between particles, the translational motion, and the vibrational frequencies. It has been argued that a more sophisticated treatment of molecular rotation via the model of a hindered rotor would lead to a considerable complication of the rrho approach as well as to a large degree of system specificity. In contrast, the effect on the vibrational entropy was found to be negligible upon application of experimentally determined frequency shifts for the transition from the gas to the liquid phase (see Sect. 3.2.3.1).
- Consequently, the proposed high density correction to the rrho model for the prediction of entropy changes either has to be accomplished in terms of the translational degrees of freedom or in terms of the incorporation of interparticle interaction. The effect of the latter point will be investigated in the next chapter of this thesis, and the present chapter is focussed on a correction of the association entropy via a more appropriate treatment of translational motion in solution.
- The free translational volume for the components of the examined supramolecular structures (wheel, axles, pseudorotaxanes) is estimated to approximately amount to $v_f = 2.5 \times 10^{-31}$ m^3 on the basis of the density of the solvent (chloroform) and a comparison between the intermolecular distance of two chloroform molecules and the corresponding distance in the chloroform complex **V6**. Comparing this estimate to the free volume per particle in the Sackur–Tetrode equation ($v_f = 4.1 \times 10^{-26}$ m^3), a scaling factor for the inverse number density of $\delta_{\rho_n^{-1}} = 6.2 \times 10^{-6}$ is obtained, and from this value a modification of the translational entropy change of $T\delta S_{\text{trans}}(\delta_{\rho_n^{-1}}) = +29.7$ kJ/mol at

$T = 298.15$ K according to Eq. 3.29. In combination with a smaller correction for the high temperature limit, this contribution results in free energy changes being more accurate than the ones predicted by the unmodified direct formation approach as expressed in Eq. 3.12, see Table 3.15.

In conclusion it is noted that entropy changes in the condensed phase obtained from the rrho approach are more accurate than one might expect due to the simplicity of the model if some care of the translational degrees of freedom is taken into account. This can either be done by setting up model reactions in which the number of participating molecules is constant during the reaction course (thereby avoiding an interconversion between different degrees of freedom and any net excess effect of the translational contributions) or by estimating the actual length scale available for translational motion in the condensed phase under examination and applying the free translational volume obtained in this way for the calculation of the translational entropy instead of the ideal gas volume applied in the Sackur–Tetrode equation. Both methods are capable of reproducing the experimentally measured free energy changes for the pseudorotaxane association in solution to approximately 10 kJ/mol and thereby give qualitatively correct predictions concerning the thermodynamic stability of these complexes. With regard to the numerous approximations finally contributing to the computational rrho approach, this result can be considered as accurate, although it is not really clear to what extent this observed accuracy benefits from system-specific error cancellations due to the chosen model systems. With regard to this point a systematic evaluation of predicted association entropies for a larger test set with regard to experimentally determined values would be helpful. According to the results and conclusions presented in this chapter, a particle-conserving model reaction as expressed in Eq. 3.2 can be suggested for association entropies in solution as the most straightforward approach avoiding any empirical estimates with regard to static quantum chemical calculations. If this is not possible for any reason, the unmodified rrho model will likely overestimate the loss of translational entropy in solution upon association as already reported in numerous studies on this subject, see the literature review sections in this chapter. As a reasonable correction scheme to overcome this problem the free volume model from the cell theory of the liquid phase can be suggested, which explicitly accounts for a reduced translational length scale at higher densities and which is not limited to a specific solvent like, e.g., water. However, the free volume v_f of a given species in solution cannot directly be calculated in terms of quantum chemical methods and therefore has to be obtained from empirical estimates like the density of the solvent or via different computational methodologies (e.g., md simulations or the application of continuum solvation models) [113–115]. However, the determination of the free volume for a solute in the condensed phase should in general be more easily accessible than the determination of its vaporization entropy, which could alternatively be employed for the correction of the computed association entropy. Thus, a completely general approach for the calculation of association entropies in solution on the basis of the rrho model being independent of any system-specific data or the stoichiometry of the underlying reaction does not seem to be possible according to the analysis presented in this chapter.

Besides a more accurate treatment of the available translational volume the neglect of interparticle interactions has been identified as a major approximation of the rrho model in the beginning of this section with regard to the treatment of the condensed phase. As pointed out in Sect. 2.2.1, an approximate treatment of these interparticle interactions on the basis of static quantum chemical calculations can be accomplished in terms of the qce approach, which can be understood as a van der Waals extension to the conventional rrho model. The performance of this model for the prediction of condensed phase entropies (and thereby the importance of interparticle interactions for this particular quantity) will be investigated in the following chapter.

References

1. Spickermann C, Felder T, Schalley CA, Kirchner B (2008) Chem Eur J 14:1216–1227
2. Reckien W, Spickermann C, Eggers M, Kirchner B (2008) Chem Phys 343:186–199
3. Kirchner B, Spickermann C, Reckien W, Schalley CA (2010) J Am Chem Soc 132:484–494
4. Schalley CA, Weilandt T, Brüggemann J, Vögtle, F (2004) Top Curr Chem 248:141–200
5. Busch DH, Stephensen NA (1990) Coord Chem Rev 100:119–154
6. Gerbeleu NV, Arion VB, Burgess J (1999) Template synthesis of macrocyclic compounds. Wiley-VCH, Weinheim
7. Diederich F, Stang PJ (2000) Templated organic synthesis. Wiley-VCH, Weinheim
8. Jäger R, Vögtle F (1997) Angew Chem Int Ed Engl 36:930–944
9. Balzani V, Venturi M, Credi A (2003) Molecular devices and machines. Wiley-VCH:Weinheim
10. Balzani V, Credi A, Silvi S, Venturi M (2006) Chem Soc Rev 35:1135–1149
11. Anelli PL, Spencer N, Stoddart JF (1991) J Am Chem Soc 113:5131–5133
12. Walker JE (1998) Angew Chem Int Ed 37:2309–2319
13. Boyer PD (1998) Angew Chem Int Ed 37:2297–2307
14. Schalley CA, Beizai K, Vögtle F (2001) Acc Chem Res 34:465–476
15. Hunter CA (1992) J Am Chem Soc 114:5303–5311
16. Pauling LJ (1954) Phys Chem 58:662–666
17. Reiher M, Sellmann D, Hess BA (2001) Theor Chem Acc 106:379–392
18. Boys SF, Bernardi F (1970) Mol Phys 19:553–566
19. Wells BH, Wilson S (1983) Chem Phys Lett 101:429–434
20. Reed AE, Curtiss LA, Weinhold F (1988) Chem Rev 88: 899–926
21. McQuarrie DA (1976) Statistical mechanics. Harper and Row, New York
22. McQuarrie DA, Simon JD (1997) Physical chemistry. University Science Books, Sausalito
23. Kirchner B, Reiher M (2007) Theoretical methods for supramolecular chemistry. In: Schalley CA (eds) Analytical methods in supramolecular chemistry. Wiley-VCH, Weinheim
24. Hirose K (2007) Determination of binding constants. In: Schalley CA (eds) Analytical methods in supramolecular chemistry. Wiley-VCH, Weinheim
25. Peter C, Oostenbrink C, van Dorp A, van Gunsteren WF (2004) J Chem Phys 120:2652–2661
26. Andersson Y, Langreth DC, Lundqvist BI (1996) Phys Rev Lett 76:102–105
27. Kamiya M, Tsuneda T, Hirao K (2002) J Chem Phys 117:6010–6015

28. Giauque WF, Kemp JD (1938) J Chem Phys 6:40–52
29. Curtiss LA, Frurip DJ, Blander M (1979) J Chem Phys 71:2703–2711
30. Frurip DJ, Curtiss LA, Blander M (1980) J Am Chem Soc 102:2610–2616
31. Reiher M, Spickermann C, Kirchner B (2010) In preparation
32. Doty P, Myers GE (1953) Discuss Faraday Soc 13:51–58
33. Doty P, Edsall JT (1951) Adv Protein Chem 6:35–121
34. Steinberg IZ, Scheraga HA (1963) J Biol Chem 238:172–181
35. Schwarzenbach G (1952) Helv Chim Acta 35:2344–2363
36. Page MI, Jencks WP (1971) Proc Nat Acad Sci USA 68:1678–1683
37. Dunitz JD (1995) Chem Biol 2:709–712
38. Doig AJ, Williams DH (1992) J Am Chem Soc 114:338–343
39. Holtzer A (1995) Biopolymers 35:595–602
40. Jencks WP (1981) Proc Nat Acad Sci USA 78:4046–4050
41. Tidor B, Karplus M (1994) J Mol Biol 238:405–414
42. Searle MS, Williams DH (1992) J Am Chem Soc 114:10690–10697
43. Hermans J, Wang L (1997) J Am Chem Soc 119:2707–2714
44. Fu A, Thiel W (2006) J Mol Struct 765:45–52
45. Firman TK, Ziegler T (2001) J Organomet Chem 635:153–164
46. Haras A, Michalak A, Rieger B, Ziegler T (2006) Organometallics 25:946–953
47. Woo TK, Blöchl PE, Ziegler T (2000) J Phys Chem A 104:121–129
48. Tuttle T, Wang D, Thiel W (2006) Organometallics 25:4504–4513
49. Jensen VR, Koley D, Jagadeesh MN, Thiel W (2005) Macromolecules 38:10266–10278
50. Vyboishchikov SF, Bühl M, Thiel W (2002) Chem Eur J 8:3962–3975
51. Goossen LJ, Koley D, Herman HL, Thiel W (2005) Organometallics 24:2398–2410
52. Cheong M, Ziegler T (2005) Organometallics 24:3053–3058
53. Vyboishchikov SF, Thiel W (2005) Chem Eur J 11:3921–3935
54. Tobisch S, Ziegler T (2004) J Am Chem Soc 126:9059–9071
55. Tobisch S, Ziegler T (2004) Organometallics 23:4077–4088
56. Zhu H, Ziegler T (2006) J Organomet Chem 691:4486–4497
57. Hristov IH, Ziegler T (2003) Organometallics 22:3513–3525
58. Timoshkin AY, Siodmiak, M, Korkin AA, Frenking G (2003) Comput Mater Sci 27:109–116
59. Wertz DH (1980) J Am Chem Soc 102:5316–5322
60. Ahlrichs R, Bär M, Häser M, Horn H, Kölmel C (1989) Chem Phys Lett 162:165–169
61. Neugebauer J, Reiher M, Kind C, Hess BA (2002) J Comput Chem 23:895–910
62. Jensen F (1999) Introduction to computational chemistry. Wiley-VCH, Chichester
63. Frisch MJ et al. (2004) Gaussian03
64. Bowman JM (1986) Acc Chem Res 19:202–208
65. Roy TK, Prasad MD (2009) J Chem Phys 131: 114102
66. Janzen J, Bartell LS (1969) J Chem Phys 50:3611–3618
67. Deraman M, Dore J, Powles J, Holloway JH, Chieux P (1985) Mol Phys 55:1351–1367
68. Pekeris CL (1934) Phys Rev 45:98–103
69. Botschwina P, Flügge J (1991) Chem Phys Lett 180:589–593
70. Atkins PW (1990) Physical chemistry. Oxford University Press, Oxford
71. Lehmann SBC, Spickermann C, Kirchner B (2009) J Chem Theory Comput 5:1640–1649
72. Lehmann SBC, Spickermann C, Kirchner B (2009) J Chem Theory Comput 5:1650–1656
73. Levine IN (2000) Quantum chemistry. Prentice Hall, New Jersey
74. Brehm G, Reiher M, Schneider S (2002) J Phys Chem A 106:12024–12034
75. Slanina Z (1991) Thermochim Acta 182:67–75
76. Slanina Z (1987) Comput Chem 11:231–234
77. National Institute of Standards and Technology, "NIST chemistry webbook", see http://webbook.nist.gov/
78. Reiher M, Salomon O, Hess BA (2001) Theor Chem Acc 107:48–55
79. Reiher M, Salomon O, Sellmann D, Hess BA (2001) Chem Eur J 7:5195–5202

80. Hoy AR, Bunker PR (1979) J Mol Spectros 74:1–8
81. Benedict WS, Gailar N, Plyler EK (1956) J Chem Phys 24:1139–1165
82. Neugebauer J, Reiher M, Kind C, Hess BA (2002) J Comput Chem 23:895–910
83. Neugebauer J, Hess BA (2003) J Chem Phys 118:7215–7225
84. Wolfram Research, Inc. (2005) Mathematica version 5.2. Wolfram Research, Inc., Champaign, Illinois
85. Helgaker T, Jørgensen P, Olsen J (2004) Molecular electronic-structure theory. Wiley-VCH, Chichester
86. Boese AD, Klopper W, Martin JML (2005) Mol Phys 103:863–876
87. Wilson EB Jr (1959) Adv Chem Phys 2:367–393
88. West W, Edwards RT (1937) J Chem Phys 5:14–21
89. Hirota E (1953) Bull Chem Soc Jpn 26:397–400
90. Pullin ADE (1960) Spectrochim Acta 16:12–24
91. Fenlon PF, Cleveland FF, Meister AG (1951) J Chem Phys 19:1561–1565
92. Yu YB, Privalov PL, Hodges RS (2001) Biophys J 81:1632–1642
93. Vallet V, Wahlgren U, Grenthe I (2003) J Am Chem Soc 125:14941–14950
94. Schlund S, Schmuck C, Engels B (2005) J Am Chem Soc 127:11115–11124
95. Hupp T, Sturm C, Janke EMB, Cabre MP, Weisz K, Engels B (2005) J Phys Chem A 109:1703–1712
96. Amzel LM (1997) Proteins 28:144–149
97. Zhou HX, Gilson MK (2009) Chem Rev 109:4092–4107
98. Hirschfelder J, Stevenson D, Eyring H (1937) J Chem Phys 5:896–912
99. Frank HS, Evans WM (1945) J Chem Phys 13:507–532
100. Henchman RH (2007) J Chem Phys 126:064504
101. Siebert X, Amzel LM (2004) Proteins 54:104–115
102. Chen J, Brooks CL III, Scheraga HA (2008) J Phys Chem B 112:242–249
103. Mammen M, Shakhnovich EI, Deutch JM, Whitesides GM (1998) J Org Chem 63:3821–3830
104. Trouton F (1884) Phil Mag 18:54–57
105. Gurney RW (1953) Ionic processes in solution. McGraw-Hill, New York
106. Kauzmann W (1959) Adv Protein Chem 14:1–63
107. Murphy KP, Xie D, Thompson KS, Amzel LM, Freire E (1994) Proteins 18:63–67
108. Mayer JE, Mayer MG (1940) Statistical mechanics. John Wiley, New York
109. Pelmenschikov V, Siegbahn PEM (2006) J Am Chem Soc 128:7466–7475
110. Rulíšek L, Jensen KP, Lundgren K, Ryde U (2006) J Comput Chem 27:1398–1414
111. Yu Z, Houk KN (2003) J Am Chem Soc 125:13825–13830
112. Lide DR (2000) Handbook of chemistry and physics. CRC Press, Boca Raton
113. Klamt A, Schüürmann G (1993) J Chem Soc Perkin Trans 2:799–805
114. Zhan CG, Chipman DM (1998) J Chem Phys 109:10543–10558
115. Tomasi J, Mennucci B, Cammi R (2005) Chem Rev 105:2999–3093

Chapter 4
Liquid Phase Thermodynamics from the Quantum Cluster Equilibrium Model

In the present chapter the effect of interparticle interactions (and excluded volume) on thermodynamic quantities of the liquid phase will be investigated in terms of the quantum cluster equilibrium model. The main focus will again be set to the calculation of condensed phase entropies, since this quantity proved to be problematic for the conventional rigid rotor harmonic oscillator approach. The results presented in the previous chapter demonstrate that if a partitioning of the atomic degrees of freedom of a molecule into molecular degrees of freedom is carried out, the (approximate) treatment of the volume available for translation is of increased significance for the translational entropy contribution. In the following, this result will be examined in more detail and related to the effect arising from interactions between the particles. These factors can be expected to be most important in the case of associated liquids, and the systems investigated are the liquid phases of water and hydrogen fluoride, both of which exhibit relatively strong intermolecular interactions in terms of hydrogen bonding.[1]

4.1 Quantum Cluster Equilibrium Computations for Associated Liquids

4.1.1 Systems Investigated and Cluster Sets

The computational treatment of associated liquids like water and hydrogen fluoride dates back to the very beginning of computational chemistry. Due to the non-negligible presence of intermolecular hydrogen bonds in liquid water, the natural choice for a computational treatment are Monte Carlo simulation and molecular dynamics simulation methods, and the first calculations for liquid water have been

[1] Parts of this chapter have already been published in Refs. [1–3].

C. Spickermann, *Entropies of Condensed Phases and Complex Systems*, Springer Theses, DOI: 10.1007/978-3-642-15736-3_4,

performed on the basis of these two methodologies [4, 5]. In spite of continuous methodological improvements in terms of e.g. polarizable force fields and the development of first principles md simulations, the difficulties in the computational treatment of liquid water with its unusual physical properties are still persistent and there is no atomistic computational model being capable to predict all relevant properties of water in a quantitative way [6–9]. A more recent example for the problems related to this particular substance is the discussion of the hydrogen bond number of a single water molecule in the liquid at ambient conditions and over the whole temperature domain at which the liquid is the thermodynamic stable phase [2, 3, 10–12]. The question about the local composition of liquid water was already investigated over one century ago by Röntgen, who stated that two different coordination patterns exist in liquid water, which undergo interconversion depending on the temperature of the liquid and that these are also responsible for the point of maximum density at approximately $T = 277$ K [13]. More recent computational examinations either rely on a prevalent occurrence of a fourfold coordination pattern (typically observed in traditional molecular dynamics simulations) or on a mainly twofold hydrogen-bonded local structure, but the issue remains a reason of scientific discussions up to now [3, 11, 14–16].

Due to the hazardous properties of pure hydrogen fluoride the experimental interest in this particular liquid has been rather small, especially concerning the determination of thermodynamic quantities. Nonetheless, there are several computational studies dealing with the liquid phase of this substance mainly relying on molecular dynamics simulation methods either in its traditional form employing (polarizable) force fields or on the basis of first principles approaches [17–22]. In most of these studies the focus has been set to larger structural aspects of the liquid phase instead of the local coordination pattern, which is not as manifolded as in the case of liquid water due to the more simple molecular structure of hydrogen fluoride. Results obtained from first principles md simulations indicate that linear staggered chains are an important structural motif in liquid hydrogen fluoride at ambient as well as supercritical conditions, but earlier experimental investigations also discuss the occurrence of cyclic species in the liquid phase. These studies point out that cooperative effects (which are generally expected to be of lesser importance in the liquid phase as compared to the gas phase) could be of increased importance in highly associated liquids like hydrogen fluoride [19, 21, 23]. However, more recent experimental studies conclude that a large fraction of the molecules in liquid hydrogen fluoride is associated in chain-like arrangements at both ambient and elevated temperatures [24, 25].

There are numerous studies in the literature concerned with the treatment of associated liquids in terms of the qce approach, the earliest of which investigating the temperature dependancy of the hydrogen bond network in the liquid phase of formamide [26]. As in the case of other computational methods suitable for the study of liquid phase systems, a substance frequently examined in terms of the qce model is liquid water, for which general thermodynamic quantities, phase transition properties of the liquid–solid and liquid–vapor phase transition, isotope effects and effects of cooperativity and dispersion contributions, as well as the temperature dependancy of the hydrogen bond number have been computed via this approach [1, 3, 15, 27–30]. Besides this extensive treatment, qce calculations have been performed for a variety of other condensed phase systems like e.g. liquid sulfur, formic acid, different alcohols, or cyclotriazane, to name but a few [31–36]. However, the liquid phase of hydrogen fluoride has not been treated in terms of the qce approach before and will thus serve as a test case for the predictability of the model in addition to the well-studied liquid phase of water for which a wealth of experimental data is available. In addition, a fully automated procedure for the adjustment of the excluded volume parameter b_{xv} and the intercluster interaction parameter a_{mf} (see Sect. 2.2.1) to fit experimental densities will be employed in order to validate the capability of the model for the quantitative calculation of condensed phase thermodynamic quantities [1, 37].

4.1.1.1 Water Cluster Sets

According to the derivation of the qce model in Sect. 2.2.1, the basic input data for a qce calculation is a cluster set and the corresponding mechanical quantities of each cluster which would also have to be available for a conventional rrho calculation, i.e., the mass m, the principal moments of inertia $I_{a,b,c}$, the harmonic frequencies $\{\nu_k\}$, and the degeneracy of the electronic ground state g_1 as well as the total (supramolecular) interaction energies, respectively (see Eq. 2.52).

There is no direct possibility to extract information about the composition of the cluster set out of the model or to estimate the cluster structures and sizes important for a reasonable treatment of the condensed phase, which means that in general an educated guess for the cluster set has to be made first on the basis of e.g. geometry optimizations, some sampling procedure, or experimental information. However, it is apparent from Eq. 2.58 that in the course of the calculation the population of each cluster species j is computed. This information can be employed for a consecutive optimization of the employed cluster set by deleting clusters which do not exceed a certain population limit, i.e., redundant structures or those not contributing significantly to the composition of the liquid phase at the chosen conditions are identified by the model. In the case of liquid water, it has been found that accurate results can already be obtained from a seven-membered cluster set containing the monomer **w1**, the dimer **w2**, cyclic structures up to the hexamer **w6**, and a cubic octamer **w8c** [15, 27]. This combination of cluster structures is illustrated in Fig. 4.1 and will be denoted as the 7w8cube cluster set from now on. It is apparent that this cluster set mainly covers twofold coordinated water molecules in flat cluster structures and that the w8cube cluster is the only bulky structure exhibiting a coordination number of three for each water monomer. Thus, with regard to the importance of the fourfold coordination pattern proposed in literature, larger inaccuracies might be expected for this cluster set in the low temperature domain of the liquid water phase [11, 16]. The (bsse-corrected) supramolecular interaction energies $\Delta E_{j,\mathrm{intra}}$ calculated for this cluster set by

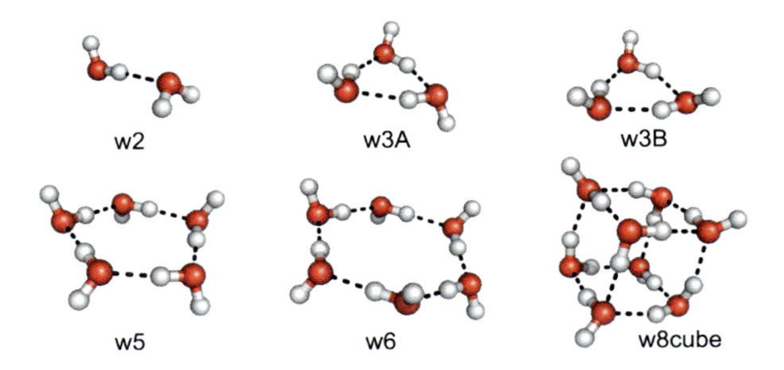

Fig. 4.1 Cluster components of the 7w8cube cluster set (without the monomer **w1**). Reprinted with permission from Ref. [2]. Copyright 2009 American Chemical Society

different quantum chemical methodologies are summarized in Table 4.1 [2, 27]. In addition to these fully cooperative energies, pair interaction energies according to

$$\Delta E_{j,\mathrm{pair}} = \sum_{k,l<k}^{i_j} E_{k,\mathrm{tot}} - E_{l,\mathrm{tot}} \tag{4.1}$$

have been computed for each cluster in the 7w8cube set, see the last column in Table 4.1. In Eq. 4.1 $E_{k,\mathrm{tot}}$ denotes the total energy of the kth monomer unit in cluster species j, and i_j labels the total number of monomer units in cluster j. In contrast to previous studies dealing with the same water cluster set in terms of the Hartree–Fock approach, the herein applied methods (Kohn–Sham density functional theory (dft) and second order Møller–Plesset perturbation theory (MP2))

Table 4.1 Supramolecular interaction energies $\Delta E_{j,\mathrm{intra}}$ (see Eq. 2.50) and the energy per monomer $\Delta E_{j,\mathrm{intra}}/i_j$ for members of the 7w8cube cluster set as obtained from different quantum chemical methods [2, 27]

Cluster	BP		B3LYP		MP2		
	TZVP	TZVPP	TZVP	TZVPP	TZVP	TZVPP	TZVPP
	Cooperative		Cooperative		Cooperative		Pair
$\Delta E_{j,\mathrm{intra}}$							$\Delta E_{j,\mathrm{pair}}$
w2	−20.9	−18.0	−22.3	−19.0	−19.9	−19.2	−19.0
w3A	−68.5	−61.4	−69.0	−61.1	−60.3	−61.7	−46.4
w3B	−63.9	−57.3	−65.2	−57.6	−57.3	−58.6	−44.0
w5	−162.3	−147.9	−161.6	−145.0	−140.4	−140.9	−68.2
w6	−201.4	−182.5	−201.0	−180.2	−175.2	−175.0	−74.3
w8cube	−311.9	−283.0	−308.5	−277.9	−269.9	−280.9	−98.4
$\Delta E_{j,\mathrm{intra}}/i_j$							$\Delta E_{j,\mathrm{pair}}/i_j$
w2	−10.5	−9.0	−11.2	−9.5	−10.0	−9.6	−9.5
w3A	−22.8	−20.5	−23.0	−20.4	−20.1	−20.6	−15.5
w3B	−21.3	−19.1	−21.7	−19.2	−19.1	−19.5	−14.7
w5	−32.5	−29.6	−32.3	−29.0	−28.0	−28.2	−13.6
w6	−33.6	−30.4	−33.5	−30.0	−29.2	−29.2	−12.4
w8cube	−39.0	−35.4	−38.6	−34.7	−33.7	−35.1	−12.3

The last column lists the pair interaction energies $\Delta E_{j,\mathrm{pair}}$ according to Eq. 4.1. All values in [kJ/mol] provide a more sophisticated treatment of electron correlation and therefore should yield more accurate data concerning cluster structures, energetics, and frequencies [15].

As can be seen from the second and third row in Table 4.1, the less symmetric trimer **w3A** is predicted to be the more stable conformer over the alternative arrangement in **w3B** by all employed methods. The values calculated for the different cluster structures adopt values in the range of approximately −20 kJ/mol for the dimer and up to approximately −310 kJ/mol for the octamer. Most stable cluster structures are predicted by the density functional theory methods with the gradient-corrected BP functional yielding the largest absolute values for almost each cluster structure. There is a considerable basis set effect visible for both the gradient-corrected BP functional and the hybrid functional B3LYP, i.e., for all investigated clusters the basis set containing the lesser number of polarization functions (TZVP) predicts more stable structures. This effect is most pronounced in the larger clusters and in the case of the octamer amounts to differences of approximately 20 kJ/mol. A reversed and less distinctive trend is apparent in the energies obtained from the MP2 calculations as well, yielding a difference of approximately 10 kJ/mol between the TZVPP and the TZVP result for the w8cube cluster. Since bsse effects have (approximately) been corrected in terms of the counterpoise correction scheme in all cases, it can be expected that these deviations are not based on technical errors alone, and a comparison of the optimized geometries shows that the hydrogen bond lengths within a given cluster for both dft methods are always larger by some pm for the TZVPP structure as those in the corresponding TZVP structure. Thus, the presence of the additional polarization functions elongates the hydrogen bond and thereby affects the cluster structure and interaction energies.

From a methodological point of view the results obtained from the MP2/TZVPP calculations can be expected to be the most reliable data, and with regard to these numbers it is apparent that most dft results overestimate the magnitude of the intracluster interaction energy. It should be noted that a comparable overbinding behavior of dft approaches has been observed previously in similar cluster calculations and also in first principles md simulations [14, 38]. The examination of the pair interaction energies listed in the last column of Table 4.1 indicates that for all clusters larger than the dimer the calculated interaction energy is poorly recovered by only considering pairwise interactions between the constituting monomer units. The deviation between the fully cooperative interaction energies and the pair interaction energies increases for the larger cluster structures, for which less than half of the cooperative interaction energies is recovered. These results demonstrate that the isolated cluster structures employed here considerably benefit from so-called *cooperative effects* (i.e., three- and higher-body contributions to the interaction energy), which additionally stabilize each hydrogen bond within a cluster due to the presence of the extra water molecules [27, 39, 40]. Such cooperative effects have been proposed to be of special importance in ring-like structures due to the donor-acceptor character of the hydrogen bond and the possibility of electron density delocalization comparable to the effect found in aromatic systems in contrast to the dipolar distribution expected for open end chain-like structures [39, 41, 42]. The second block of Table 4.1 estimates these contributions in terms of fractional interaction energies per monomer unit in the cluster. From the pairwise $\Delta E_{j,\mathrm{pair}}$ values per monomer it is apparent that the exclusive consideration of pair interactions within a cluster yields an almost constant contribution of approximately -10 to -15 kJ/mol per monomer unit to the overall pair interaction energy, i.e., there is an approximately linear relation between the pairwise interaction energy and the number of monomer units. This is in contrast to the trend found for the supramolecular interaction energies $\Delta E_{j,\mathrm{intra}}$ per monomer. For all clusters larger than the dimer the predicted energy fractions are lower by at least 5 kJ/mol as compared to the corresponding pair energy fractions. In addition, the supramolecular interaction energies do not fall off in a linear fashion, i.e., the magnitudes of the corresponding energy fractions per monomer increase with growing cluster size, thereby showing a typical behavior of cooperative systems. Due to the relatively large number of hydrogen bond contacts the most significant stabilization in terms of cooperative effects is found for the w8cube cluster regardless of the applied methodology. However, the largest *increase* in the stabilization of about 10 kJ/mol is found for the transition from the dimer to the trimer clusters. This observation can be rationalized in terms of the reduced number of hydrogen bond contacts with respect to the monomer units in the chain-like **w2** structure (two monomers, one hydrogen bond) as compared to the situation in ring clusters (*n* monomers, *n* hydrogen bonds). An almost equally large stabilization can be observed for the transition from the trimers to the **w5** cluster, but here two monomer units are added and additional effects arising from the lesser ring strain in the larger cyclic structures can be expected. For the transition from the pentamer to the hexamer a negligible stabilization of approximately 1 kJ/mol per monomer is predicted by all methods, thereby indicating that either no significant differences in the cooperative effects between those clusters are present or a possible increased stabilization due to cooperativity is counterbalanced by e.g. a spatial arrangement of the monomers more unfavorable in terms of hydrogen bonding as compared to the pentamer. Finally, the octamer again benefits from a more favorable number of hydrogen bonds per monomer unit as can be seen from Fig. 4.1 and therefore exhibits the largest stabilization per monomer of all examined structures. However, due to the magnitude of this fraction (1.5 hydrogen bonds per monomer) the computed stabilization is relatively small compared to the pentamer and hexamer structure, and the consideration of interaction energy fractions per *hydrogen bond* ($\Delta E_{\mathrm{w8cube,intra}}/n_{\mathrm{w8cube}} = -23.4\,\mathrm{kJ/mol}$, where $n_{\mathrm{w8cube}} = 12$ denotes the number of hydrogen bonds) indicates that largest cooperative effects are to be expected for the cyclic **w5** and **w6** clusters [2, 15]. It is clear that these considerable cooperative effects are a direct consequence of the isolated cluster calculations, in which any environmental effects are absent. This can e.g. be seen by comparing the values listed in Table 4.1 to supramolecular interaction energies obtained from calculations in which the electrostatic effect of the solvent is considered in terms of a continuum solvation model [43]. These values are less affected by cooperative effects and lie in the range of the pairwise interaction energies presented in Table 4.1, but such approaches only consider the electrostatic continuum effect (which is a pairwise contribution) and neglect the non-classical character of hydrogen bonding from solvent molecules outside the cluster [42, 44]. Thus, the degree of cooperativity which is present in the real liquid water phase is difficult to quantify from these cluster calculations alone.

The importance of a tetrahedral coordination pattern for the reproduction of the anomalies of liquid water as well as the general treatment of the low temperature domain including the liquid–solid phase transition has been pointed out by several authors [11, 16, 28]. In view of this reasoning additional water clusters including fourfold hydrogen bonded water molecules have been calculated to complement the previously introduced 7w8cube cluster set. These new cluster structures are illustrated in Fig. 4.2.

Due to the structural similarity of the central fourfold coordinated water molecule to the corresponding organic compounds these additional clusters are denoted as spiro clusters. The supramolecular interaction energies calculated for these new structures are summarized in Table 4.2. It is apparent that all spiro structures are predicted to be more stable than the members of the 7w8cube set with the exception of the w8cube itself, which is based on the larger number of hydrogen bond contacts in these structures. As in the case of the smaller structures from the 7w8cube set, largest absolute values for the cooperative interaction energies are obtained from the dft methods with differences to the MP2 results as high as approximately 55 kJ/mol in the case of the **s13** cluster and the TZVP basis set. For the dft methods the additional polarization functions of the TZVPP basis set again result in a destabilizing effect on the interaction energies of all spiro clusters, which amounts up to approximately 45 kJ/mol for the largest structures. This pronounced basis set effect is not seen in the case of the MP2 computations, where only the largest cluster is predicted to be less stable by the larger basis set and a reverse trend is found for the **s7** cluster, which is thereby in line with the MP2 basis set trend observed in the interaction energies of the 7w8cube cluster set.

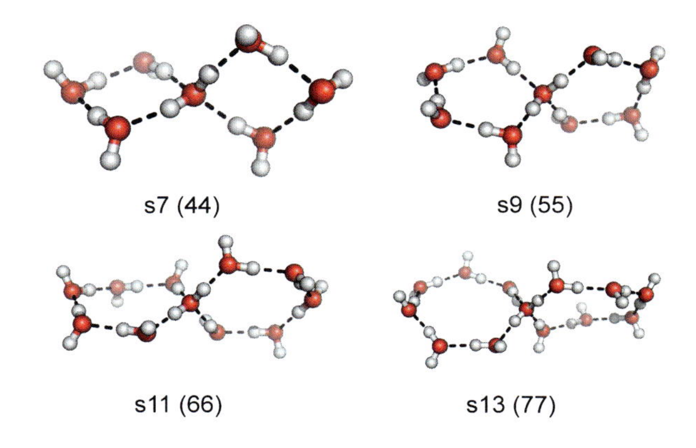

Fig. 4.2 Spiro cluster structures containing the fourfold coordination pattern. Reprinted with permission from Ref. [2]. Copyright 2009 American Chemical Society

Table 4.2 Supramolecular interaction energies $\Delta E_{j,\text{intra}}$ (see Eq. 2.50) and the energy per monomer $\Delta E_{j,\text{intra}}/i_j$ for the spiro cluster structures depicted in Fig. 4.2 as obtained from different quantum chemical methods [2, 27]

Cluster	BP		B3LYP		MP2		
	TZVP Cooperative	TZVPP	TZVP Cooperative	TZVPP	TZVP Cooperative	TZVPP	TZVPP Pair
$\Delta E_{j,\text{intra}}$							$\Delta E_{j,\text{pair}}$
s7	−247.3	−218.6	−236.5	−213.7	−203.1	−208.8	−90.6
s9	−324.7	−296.5	−321.7	−290.3	−279.9	−281.7	−75.0
s11	−392.5	−356.3	−391.8	−351.9	−342.5	−342.8	−66.5
s13	−461.6	−416.6	−462.4	−413.6	−405.2	−389.1	−52.1
$\Delta E_{j,\text{intra}}/i_j$							$\Delta E_{j,\text{pair}}/i_j$
s7	−35.3	−31.2	−33.8	−30.5	−29.0	−29.8	−13.0
s9	−36.1	−32.9	−35.7	−32.3	−31.1	−31.3	−8.3
s11	−35.7	−32.4	−35.6	−32.0	−31.1	−31.2	−6.0
s13	−35.5	−32.1	−35.6	−31.8	−31.2	−29.9	−4.0

The last column lists the pair interaction energies $\Delta E_{j,\text{pair}}$ according to Eq. 4.1. All values in [kJ/mol]

The pair interaction energies obtained for the spiro clusters are again less than half the magnitude of the fully cooperative energies, and in the case of the **s11** and **s13** structures even less than one quarter of the corresponding $\Delta E_{j,\text{intra}}$ values. Thus, the magnitude of the pair interaction energies decreases with increasing cluster size, whereas a reversed trend could be observed for the clusters of the 7w8cube set. In combination with the increasing magnitude of the supramolecular interaction energies, this is a clear indication for the elevated importance of cooperative effects in these spiro structures.

This observation is also reflected in the magnitudes of the interaction energies per monomer listed in the second block of Table 4.2. With the exception of the w8cube structure, all members of the 7w8cube set show smaller absolute $\Delta E_{j,\text{intra}}/i_j$ values, which indicates that the average stabilization of the individual monomers is relatively large in the spiro structures. With regard to this point it is also apparent that there are no large differences between the individual spiro clusters, i.e., for the stabilization of the individual monomer it does not matter if it is bound in the **s7** or in the **s13** structure. This is not necessarily to be expected according to the differences in ring size (and therefore in ring strain) between the distinct spiro clusters, see Fig. 4.2. If again the supramolecular interaction energies $\Delta E_{j,\text{intra}}$ per *hydrogen bond* are considered instead, values between −26.1 kJ/mol (**s7**) and −28.8 kJ/mol (**s13**) are calculated [2]. Compared to the corresponding numbers of the 7w8cube set, these numbers indicate a significant stabilization of each hydrogen bond in the larger cyclic clusters (**w5, w6, s11, s13**) as compared to e.g. the bulky w8cube structure, and the largest stabilization in these terms is predicted for the **w6** cluster, see Table 4.1 The direct transferability of these energetic criteria to the treatment of the liquid phase is questionable, but the importance of five- and six-membered rings for the liquid phase of water has been explicitly discussed before [15, 42].

4.1.1.2 Hydrogen Fluoride Cluster Set

As noted above, several experimental as well as computational studies discuss the occurrence of staggered chains of hydrogen-bonded molecules in the liquid phase of hydrogen fluoride, but there are also studies emphasizing the possible importance of ring structures in the liquid [22–24, 45–47]. Some of the computational investigations provide estimates for the average number of monomer units in hydrogen fluoride chains and rings to be equal or less than eight [19, 22]. With regard to these observations quantum chemical calculations have been performed for several possible chain- and ring-like arrangements of hydrogen fluoride monomer units employing dft methods as well as second order Møller–Plesset perturbation theory in combination with basis sets of triple-ζ (TZVP) and quadruple-ζ (QZVP) quality, respectively. An overview of all considered cluster structures is given in Fig. 4.3. However, the results obtained from the MP2 normal mode analysis show that chains including more than three monomer units in general yield one (or more) negative frequency values and therefore are no minima on the MP2 potential energy surface (with the exception of the MP2/TZVP calculation), see Sect. 7.1 [38]. Thus, the only open-end chain structures present in all hydrogen fluoride cluster sets are the dimer **c2** and the trimer chain **c3c**, and the inclusion of the hexamer chain structure **c6c** is only realized in the case of the dft calculations. For the calculated MP2/QZVP structures additional geometry optimizations and normal mode analyses employing a smaller (i.e., more precise) convergency criterion for the norm of the cartesian gradient have been carried out (see Sect. 7.1) in order to obtain an accurate set of geometries. These refined structures have been subsequently employed for the calculation of high quality ab initio interaction energies and complete basis set (cbs) limit extrapolation studies in terms of the coupled cluster method including single and double excitations and the perturbative treatment of triple excitations (CCSD(T)) as well as the incremental scheme in case of the larger cluster structures [48–50]. The calculated interaction energies from all considered methods are summarized in Table 4.3 (the column "QZVP*" lists results based on the more accurate geometry optimizations). It is apparent that in the case of the hydrogen fluoride dimer **c2** the calculated values are in the same order of magnitude as in the case of the water dimer (approximately -19 kJ/mol), which indicates that the isolated hydrogen bond in both substances is of comparable strength. The interaction energies obtained from the PBE functional in combination with the TZVP basis set yield more stable clusters than any of the other approaches no matter which of the clusters is considered. In the case of the dimer, this overbinding amounts to approximately 2 kJ/mol as compared to the CCSD(T)/cbs result and reaches a difference of nearly 50 kJ/mol in the case of the cyclic octamer **c8r**.

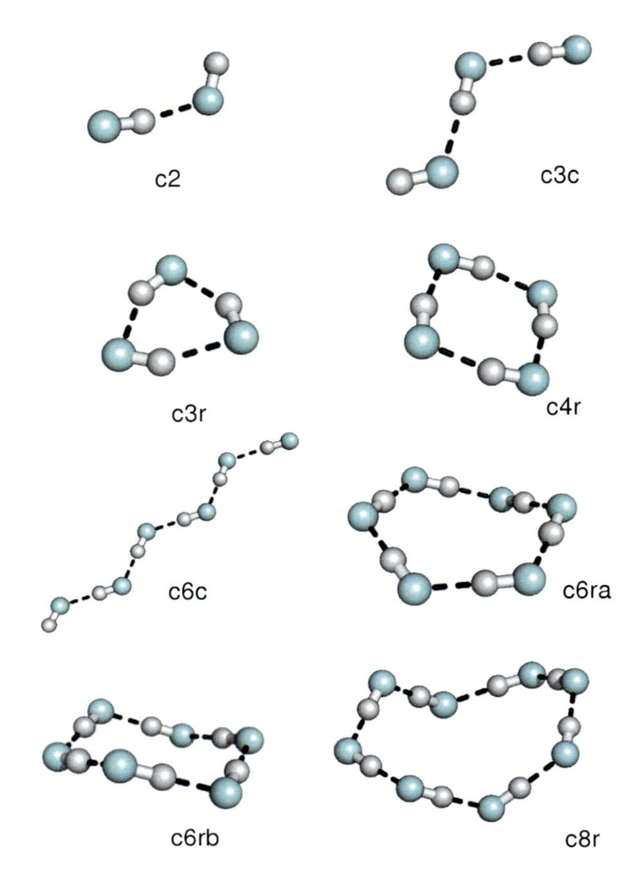

Fig. 4.3 Cluster components of the hydrogen fluoride cluster set (without the monomer **c1**)

This tendency of gradient-corrected dft has also been observed in the calculations of the water clusters (see the first two columns in Table 4.1), but not to such an extent. In contrast, the second gradient-corrected density functional (BP) shows no uniform trend compared to the interaction energies obtained from the CCSD(T)/cbs approach. Beginning with the cyclic tetramer **c4r**, all larger clusters are predicted to be more stable, whereas the smaller clusters show lesser magnitudes of the supramolecular interaction energy with respect to the CCSD(T)/cbs reference. However, this underbinding tendency is of insignificant extent (less than or equal 2 kJ/mol in all cases), while the overbinding tendency found for the larger clusters assumes differences of approximately 20 kJ/mol in the case of the octamer cluster.

In the case of the MP2 results the basis set effect already found for the water clusters in Tables 4.1 and 4.2 is even more pronounced as can be seen from the central block in Table 4.3. For the transition from the TZVP to the QZVP basis set a lowering in interaction energy of more than 20 kJ/mol is predicted in the case of the **c8r** cluster, and an additional decrease again of approximately 20 kJ/mol is

Table 4.3 Supramolecular interaction energies $\Delta E_{j,\mathrm{intra}}$ (see Eq. 2.50) and the energy per monomer $\Delta E_{j,\mathrm{intra}}/i_j$ for the hydrogen fluoride cluster structures depicted in Fig. 4.3 as obtained from different quantum chemical methods [50]

Cluster	CCSD(T)	MP2				BP	PBE
	CBS	CBS	QZVP$^{☆}$	QZVP	TZVP	TZVP	TZVP
$\Delta E_{j,\mathrm{intra}}$							
c2	−19.5	−19.2	−17.8	−17.8	−17.0	−17.9	−21.4
c3c	−45.1	−44.6	−41.5	−41.5	−39.3	−43.1	−50.1
c3r	−65.1	−64.0	−59.2	−59.1	−50.5	−64.4	−72.6
c4r	−119.0	−118.4	−109.8	−109.8	−95.0	−124.5	−136.5
c6c	–	–	–	–	–	−137.9	−154.9
c6r$_a$	−198.7	−198.9	−184.3	−184.3	−165.1	−212.3	−231.7
c6r$_b$	−199.1	−199.2	−184.6	−184.6	−165.2	−212.9	−232.4
c8r	−268.8	−269.0	−249.4	−249.0	−225.1	−288.9	−314.4
$\Delta E_{j,\mathrm{intra}}/i_j$							
c2	−9.7	−9.6	−8.9	−8.9	−8.5	−9.0	−10.7
c3c	−15.0	−14.9	−13.8	−13.8	−13.1	−14.4	−16.7
c3r	−21.7	−21.3	−19.7	−19.7	−16.8	−21.5	−24.2
c4r	−29.8	−29.6	−27.4	−27.4	−23.8	−31.1	−34.1
c6c	–	–	–	–	–	−23.0	−25.8
c6r$_a$	−33.1	−33.1	−30.7	−30.7	−27.5	−35.4	−38.6
c6r$_b$	−33.2	−33.2	−30.8	−30.8	−27.5	−35.5	−38.7
c8r	−33.6	−33.6	−31.2	−31.1	−28.1	−36.1	−39.3

The column "QZVP$^{☆}$" denotes results obtained from a more accurate convergency criterion for the geometry optimization (see Sect. 7.1). All values in [kJ/mol]

observed for the extrapolation to the cbs limit. For most smaller clusters it is found that the interaction energies obtained from the MP2/QZVP combination are closer to the ones predicted for the cbs limit than to the values of the MP2/TZVP combination, i.e., the gain in accuracy for the change from the TZVP set to the QZVP is comparatively large in these clusters at least concerning the computed energies.

It is also apparent that the effect of increasing the accuracy of the geometry optimization on the MP2/QZVP interaction energies is completely negligible and in the case of the smaller clusters virtually leads to no difference between the computed numbers, see the column labeled "QZVP$^{☆}$". The comparison between the dft and MP2 methods at the same basis set (TZVP) again shows large discrepancies between the two approaches, in which the dft methods always predict more stable cluster structures with differences as large as approximately 90 kJ/mol in the case of the **c8r** structure. However, these discrepancies are smaller if the basis set size for the MP2 calculations is increased, and in the cbs limit of the MP2 interaction energies the differences to the BP/TZVP results do not exceed approximately 20 kJ/mol. It is interesting to see that the open end chain clusters **c2** and **c3c** are energetically disfavored by the BP functional (predicted to be less stable as compared to MP2/cbs), whereas all cyclic species exhibit larger magnitudes in the interaction energy as compared to MP2/cbs. However, the size of the sample is definitely too small to arrive at a general conclusion in that direction. Comparing the MP2/cbs energies to the corresponding CCSD(T)/cbs values shows that both approaches agree to a large extent in all examined cases with differences smaller than 1 kJ/mol for almost all cluster structures. In the case of the smaller systems the coupled cluster approach predicts slightly more stable structures, whereas a reversed trend is found for the larger rings $\mathbf{c6r}_a$, $\mathbf{c6r}_b$, and **c8r**. A comparison between the calculated dimer interaction energies listed in the first row of Table 4.3 and an experimental measurement of the hydrogen fluoride dimer dissociation energy ($D_e = 19.1(+1.2, -1.1)$ kJ/mol) demonstrates that the MP2/cbs and the CCSD(T)/cbs approaches are the most accurate methods employed in this study [51].

The interaction energies per monomer unit $\Delta E_{j,\mathrm{intra}}/i_j$ are listed in the second block of Table 4.3. According to these numbers, largest stabilization of the monomer units is found for the large ring clusters and the dft methods as to be expected from the observed trends in the supramolecular interaction energies. In the case of the trimer structures a considerably larger stabilization is obtained for the ring isomer due to the additional hydrogen bond. However, if this contribution is estimated by adding the interaction energy of an isolated hydrogen bond (obtained from the **c2** cluster) to the corresponding value of the chain isomer **c3c**, it is seen that there is virtually no preference for the ring structure predicted by any of the employed methods. This observation can again either be rationalized by the absence of significant differences between cooperative effects in the hydrogen bonds of the **c3c** and the **c3r** cluster, respectively, or by a compensating effect due to ring strain in the small cyclic cluster **c3r**. However, the relatively large stabilization of approximately 10 kJ/mol per monomer predicted by all employed methods for the transition from the cyclic trimer **c3r** to the cyclic tetramer **c4r** and the larger ring structures indicates that a considerable release of ring strain upon this transition is very likely. A more distinguished analysis of cooperative effects in chain and ring structures can be done in terms of a comparison between the hexamer chain and ring isomers as obtained from the dft methods (due to the higher order saddle point characteristic of the **c6c** structure on the MP2/QZVP potential energy surface this method is not considered in that regard). For both density functionals the difference in interaction energy per monomer unit amounts to approximately 13 kJ/mol in favor of the cyclic species $\mathbf{c6r}_a$ and $\mathbf{c6r}_b$. If this number is compared to the stabilization obtained from an additional (isolated) hydrogen bond per monomer (in a hexamer this amounts to approximately 3 kJ/mol), it is clear that the larger ring clusters benefit from increased cooperative effects as compared to the ones found in the hexamer chain **c6c**.

Additional insight into the degree of cooperativity of the different hydrogen fluoride clusters can be obtained by analyzing the electron occupancy of certain orbitals at the acceptor units of the intracluster hydrogen bond. In general, the canonical scf orbitals are delocalized over the whole molecular structure and therefore cannot be unambiguously assigned to a certain atom or group of atoms.

In order to perform a reasonable population analysis with regard to hydrogen bonding, the canonical scf orbitals have to be localized first. In the present case the natural bond orbital (NBO) localization followed by a natural population analysis (NPA) has been employed [41, 42, 52]. In the localized natural orbital basis the occupancy of the $\sigma^{\star}$ antibonding orbital of the intramolecular hydrogen fluoride bond can provide information about the amount of charge transfer between distinct hydrogen fluoride units, because this orbital is unoccupied in the non-interacting hydrogen fluoride monomer unit. The average occupancy of the $\sigma^{\star}$ orbital as obtained from the NPA is illustrated in Fig. 4.4 as a function of the cluster size and the cluster type (chain or ring).

From this plot it is apparent that the amount of charge transfer due to hydrogen bonding is always larger in the case of the ring clusters regardless of the cluster size or the electronic structure method applied. In most cases the degree of occupation obtained for the chain clusters is equal or less than half the value obtained from the same method for the ring cluster of corresponding size. In addition, the dotted lines indicate an almost perfectly linear relation between the number of monomers i_j in a chain cluster and the corresponding average occupancy of the $\sigma^{\star}$ orbital, i.e., the occupancy of a chain cluster containing n monomer units can be approximated by $n - 1$ times the occupancy of the dimer **c2** to good precision. In contrast, the occupancies calculated for the cyclic clusters show a considerable increase for the transition from the trimer to the tetramer and from the tetramer to the hexamer, after which the occupancy remains approximately constant. If the dimer is additionally taken into account, the largest

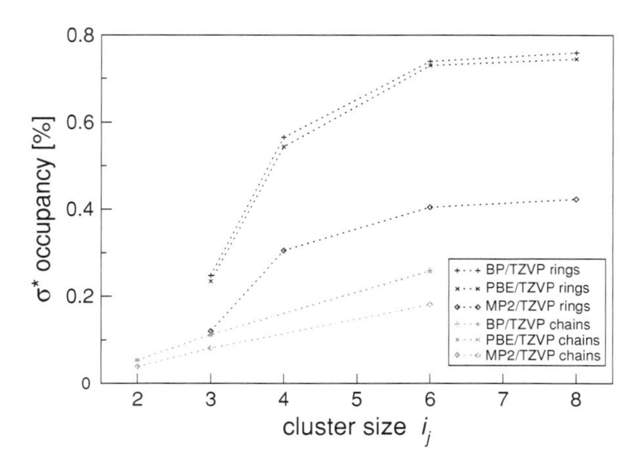

Fig. 4.4 Average occupancy of the $\sigma^{\star}$ orbital of the H–F bond as a function of the cluster type [50]. For $i_j = 6$ the boat-like conformer $\mathbf{c6r}_a$ has been employed in the case of the cyclic clusters. The dotted lines are meant to guide the eye. Please note that the hexamer chain cluster **c6c** is a minimum on the MP2/TZVP potential energy surface and therefore is employed in this analysis as well. However, in order to ensure comparability to the remaining post-Hartree–Fock methods this structure is not considered in the qce results presented in this chapter. This is also the reason why no interaction energy for this cluster is listed in Table 4.3

cooperative contributions are obtained for the trimer and the tetramer (increase in occupation by a factor of approximately 5 for **c2** → **c3r** and 2 for **c3r** → **c4r**, respectively, in the case of the dft results), which is in accurate agreement to the results obtained in a previous study [53]. Accordingly, the almost linear dependancy of the $\sigma^{\star}$ occupation on the cluster size predicted for the chain structures is absent and an approximately logarithmic relationship can be estimated, which only leads to a relatively small increase in cooperative effects at larger cluster sizes. This result can also help to explain why the majority of ringlike structures found in first principles md simulations of liquid hydrogen fluoride contains up to eight monomer units, because the gain in cooperativity for even larger aggregates is rather small and the loss of entropy upon ring closure could be more severe, thereby favoring the formation of chain structures of larger size [19, 22]. As in the case of the supramolecular interaction energies listed in Table 4.3 the occupancies obtained from the MP2/TZVP NBO analysis are smaller by a factor of approximately 0.5 in all cases than the ones predicted by the dft methods, which yield almost identical $\sigma^{\star}$ occupations for all clusters. This in turn indicates that the occupancies cannot directly be correlated to the supramolecular interaction energies, since the energies computed by the two considered density functionals differ significantly as can be seen in the last two columns of Table 4.3. However, it should be kept in mind that the supramolecular interaction energy is the difference between the *total* energies of the cluster and the relaxed monomer units and therefore includes all possible interactions reasonably described by the applied quantum chemical method, whereas the occupancy of the $\sigma^{\star}$ orbital is a measure for the hydrogen bond interaction only and does not cover additional interactions e.g. between permanent molecular multipoles.

4.1.2 Zeroth Order Quantum Cluster Equilibrium Calculations

As pointed out in Sect. 2.2.1 the quantum cluster equilibrium approach to the condensed phase involves the determination of two empirical parameters, namely the intercluster interaction parameter a_{mf} (see Eq. 2.51) and the excluded volume factor b_{xv} according to Eq. 2.48. These parameters control the van der Waals-like contributions from the volume of the particles as well as the mean field interaction between different clusters, but as pointed out in the discussion of Eq. 2.45 a certain fraction of the *intermolecular* interactions is captured in the first principles cluster calculations as *intracluster* interactions. Thus, in order to examine the effect of this partial treatment of intermolecular interactions, the qce model will first be employed in a zeroth order setup, in which the residual intercluster interaction is not considered ($a_{\mathrm{mf}} = 0$ (J × m^3)/mol) and the volume of the particles is either included as the unmodified sum of the atomic van der Waals volumes ($b_{\mathrm{xv}} = 1$) or completely neglected ($b_{\mathrm{xv}} = 0$). The water and hydrogen fluoride cluster sets introduced in the previous section will serve as the model systems for these investigations.

4.1.2.1 Complete Neglect of Cluster Volume

In order to obtain an understanding of the exclusive influence of the transition from the single molecule approach routinely applied in many static quantum chemical calculations to the cluster approach of partly associated molecules, the contribution of the reduced volume effect due to the cluster volume will be ignored first.

Besides the external pressure p and the temperature T several technical parameters have to be provided for a qce calculation like for instance the maximum number of iterations n_{cyc} or the convergence criterion V_{crit} for the computed phase volume. A summary of these details and the values assigned to the corresponding parameters can be found in Sect. 7.1. The first system examined in terms of this complete parameter-free setup will be the liquid phase of water in terms of the 7w8cube cluster set illustrated in Fig. 4.1 and the extended cluster set additionally including the spiro clusters (7w8cube + spiro clusters, see Fig. 4.2). The molar volume as a function of the temperature calculated on the basis of this parameter setup and of the cluster properties obtained from the different quantum chemical methods considered is plotted in Fig. 4.5. From the curves in Fig. 4.5 it is apparent that all methods predict an ideal gas-like behavior for the cluster equilibrium at temperatures larger than $T = 280$ K. Below this temperature a volume collapse can be observed in all curves obtained from fully cooperative interaction energies, whereas the isobar based on the pairwise MP2/TZVPP interaction energies from Table 4.1 retains the ideal gas character over almost the whole investigated temperature domain. Due to the smooth decline of the curves the calculated condensation processes do not correspond to real first order phase transitions, but nevertheless represent a decrease in molar volume to approximately 15% of the corresponding ideal gas volume over a

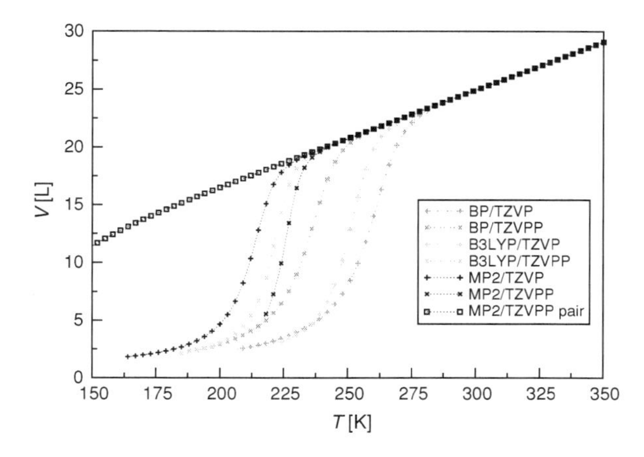

Fig. 4.5 Molar volume V as a function of the temperature T as obtained from different quantum chemical methods employed for the calculation of the 7w8cube cluster set. The lowest temperature of each isobar corresponds to the point at which the qce iterations no longer converge (maximum number of iterations $n_{cyc} = 1{,}000$). All calculations refer to a pressure of $p = 1 \times 10^5$ Pa

relatively small temperature interval of approximately 20 K. There are considerable differences in the onset of the condensation process predicted by the various quantum chemical methods. The highest condensation temperature of approximately $T = 280$ K is obtained for the BP/TZVP curve, followed by the B3LYP/TZVP isobar (approximately $T = 270$ K). Besides the difference in the transition temperature, these two isobars also differ in the slope of the curve in the condensation interval with the B3LYP isobar showing a steeper decline. The condensation temperature of the curves obtained from the dft methods in combination with the larger TZVPP basis set are shifted to lower temperatures by approximately 30 K (BP) and 50 K (B3LYP), respectively, but at the same time their characteristic slope in the condensation interval is retained. In the case of the MP2 cluster calculations, the obtained isobars exhibit a considerable shift of the condensation temperature towards lower values ($T = 240$ K for MP2/TZVPP and $T = 225$ K for MP2/TZVP, respectively) as compared to the dft isobars. In addition, the absolute difference in the condensation point between the TZVP and the TZVPP basis sets is smaller as in the case of the dft curves, and the decline of the MP2/TZVPP isobar is even more pronounced as the one found for the B3LYP/TZVPP curve. However, the major difference between the isobars obtained from the MP2 and dft cluster set calculations clearly lies in the reversed basis set trend with regard to the condensation temperature. In the case of the MP2 calculations the isobar based on the larger basis set exhibits a higher condensation temperature as the corresponding TZVP curve, whereas the opposite trend is observed for the dft calculations. This reversed basis set effect of the MP2 and dft results has been found previously in the supramolecular interaction energies listed in Table 4.1, for which an additional stabilization of the cluster structures for the larger basis set in the case of the MP2 approach and the opposite trend in the case of the dft methods is observed. Consequently, it is reasonable to expect that larger magnitudes in the interaction energy lead to an increase of the transition temperature between the low density ideal gas phase and the high density liquid-like phase. This conclusion is additionally supported by the fact that the sequence of phase transition points is almost perfectly matched by the corresponding order of stability in terms of the supramolecular interaction energies (with the exception of the MP2/TZVPP and the B3LYP/TZVPP results). The most stable cluster structures (obtained from the BP/TZVP combination, see Table 4.1) also exhibit the highest phase transition temperature, and the ideal gas phase of the least stable set (MP2/TZVP) covers the largest temperature domain of all investigated methods, i.e., it does exist at temperatures for which all remaining ideal gas phases are no longer stable. Furthermore, the isobar obtained from the MP2/TZVPP pair interaction energies shows no transition to a high density phase at all in the investigated temperature interval, which is in accordance to the fact that the magnitudes of these energies are significantly smaller than those of the fully cooperative energies, regardless of the applied quantum chemical method. These findings also have been reported in previous qce studies on liquid water for a zeroth order parameter setup in which the intercluster interactions are equally switched off as in the present case ($a_{\mathrm{mf}} = 0$ (J $\times$ m^3)/mol), but the excluded volume has been considered in an unmodified fashion ($b_{\mathrm{xv}} = 1$) [15, 27].

From the results presented here it is clear that this shift of the phase transition point is directly based on the stability differences between the clusters as obtained from different interaction energies and no volume effect possibly arising from varying V_{ex} values calculated for different cluster equilibria in terms of different quantum chemical methodologies.[2] This observation can also be expected according to the physical reasoning that at a given temperature stronger intermolecular interactions (as expressed in more stable cluster structures) tend to increase the density of the system under investigation and require larger amounts of energy for the evaporation as indicated in increased vaporization enthalpies and higher boiling points (for which liquid water is a demonstrative example) [54, 55]. Thus, the simple model of an ideal cluster gas as introduced in Sect. 2.2.1 is capable of capturing the essential features of the phase transition from the low density gas phase to a high density liquid-like phase in a qualitative way. The extension of the 7w8cube cluster set by the use of the spiro clusters illustrated in Fig. 4.2 introduces the fourfold coordination pattern into the cluster set. The isobars calculated for this extended cluster set in the complete parameter-free qce setup ($a_{mf} = 0$ (J $\times$ m^3)/mol, $b_{xv} = 0$) are plotted in Fig. 4.6. The isobars in Fig. 4.6 indicate that the additional spiro clusters do not affect the general progression of the different methods over the examined temperature domain and that the onset of the condensation process is also only barely affected by the extended cluster set. However, it is apparent that in the present case all isobars converge to the lowest temperature considered in the qce calculation (with the exception of the B3LYP/TZVP curve showing some unconverged iterations in the phase transition part), which is not the case if the 7w8cube cluster set is employed exclusively (compare the low temperature part of Figs. 4.5 and 4.6). From this observation it can be inferred that the fourfold coordination pattern as introduced via the spiro clusters is an important structural element in the low temperature regime of liquid water and that this motif is necessary for a consistent modeling of this part of the water phase diagram in terms of the (zeroth order) qce approach [3, 16]. In addition, this result again indicates that the partial treatment of intermolecular interactions via the isolated cluster approach is sensitive enough to detect the significance of this coordination pattern. The isobars calculated for the hydrogen fluoride cluster set (see Fig. 4.3) and the parameter-free qce approach are plotted in Fig. 4.7. It is apparent that the obtained curves qualitatively agree to the ones calculated for the extended water set (see Fig. 4.6), but the condensation temperatures are shifted to larger values. In addition, all curves converge down to $T = 150$ K as in the case of the extended water cluster set including the spiro clusters, thereby indicating that no important coordination pattern is missing as in the low temperature domain of the 7w8cube water cluster set. The direct correlation between the strength of intermolecular interaction, i.e., the magnitude of the

[2] According to Eq. 2.48, the excluded volume contribution depends on the calculated cluster populations, which in turn are related to the applied methodology via the cluster partition functions $\{q_j\}$ as expressed in Eq. 2.58.

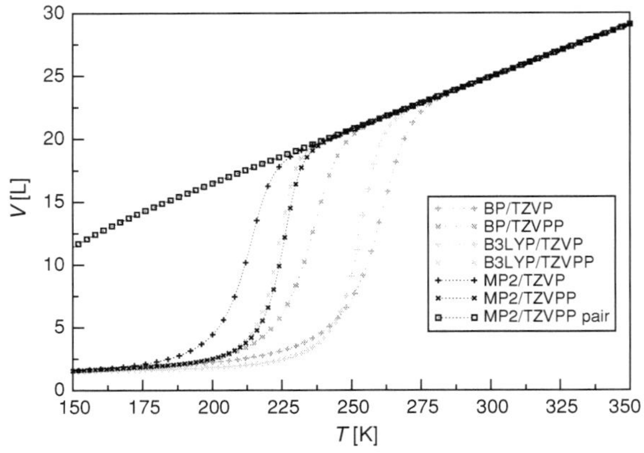

Fig. 4.6 Molar volume V as a function of the temperature T as obtained from different quantum chemical methods employed for the calculation of the extended (7w8cube + spiro) cluster set. The lowest temperature of each isobar corresponds to the point at which the qce iterations no longer converge (maximum number of iterations $n_{\mathrm{cyc}} = 1{,}000$). All calculations refer to a pressure of $p = 1 \times 10^5$ Pa

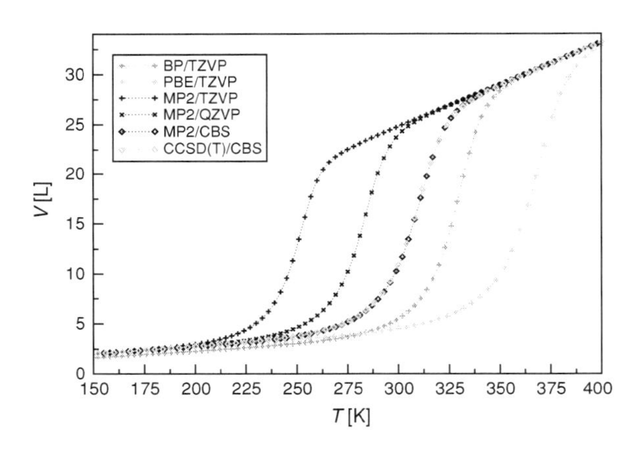

Fig. 4.7 Molar volume V as a function of the temperature T as obtained from different quantum chemical methods employed for the calculation of the hydrogen fluoride cluster set (see Fig. 4.3). The lowest temperature of each isobar corresponds to the point at which the qce iterations no longer converge (maximum number of iterations $n_{\mathrm{cyc}} = 1{,}000$). All calculations refer to a pressure of $p = 1 \times 10^5$ Pa

supramolecular interaction energies (see Table 4.3), and the onset of the condensation process found for the water cluster sets is again clearly visible in Fig. 4.7. Both density functional methods predict the most stable cluster structures and accordingly the corresponding isobars show a stable high density phase over the largest temperature interval. In contrast, the MP2 calculations with the triple-ζ and quadruple-ζ basis sets yield less stable cluster structures and consequently the condensation temperatures are shifted to lower values, whereas the curves obtained from the complete basis set extrapolations are intermediate both regarding the magnitude of the interaction energies as well as the condensation temperatures. Furthermore, the comparison between the interaction energies and isobars obtained from the MP2/cbs and the BP/TZVP combinations indicates that the point of condensation is more significantly affected by the energetics of the larger cluster structures, since in the case of the smaller clusters the BP/TZVP approach predicts less stable clusters and therefore should also lead to a lower condensation temperature. This increased contribution of the hexamers and the octamer is also reflected in the population computed at each temperature point of the qce calculation according to Eq. 2.58 (not depicted). The apparent equality of the MP2/cbs and CCSD(T)/cbs isobars furthermore show that the parameter-free qce model is not sensitive to differences in the cluster interaction energy which are below 1 kJ/mol. As pointed out above, the transition from the ideal gas phase to the high density liquid-like phase occurs at higher temperatures in the case of the hydrogen fluoride cluster set. It is therefore instructive to compare the calculated condensation temperatures to the experimental boiling point of liquid hydrogen fluoride, which for a pressure of $p = 1 \times 10^5$ Pa occurs at $T = 292.69$ K [56]. With regard to this reference value, it is visible from Fig. 4.7 that the isobars obtained from the complete basis set limits of the MP2 and CCSD(T) methods as well as the MP2/QZVP curve are most accurate in predicting an onset of the transition to the gas phase at approximately 290 and 260 K, respectively (see Fig. 5.3 for a more precise determination of the phase transition temperatures via the turning points in the specific heat capacities). This is in accordance with the fact that these two methods (MP2 and CCSD(T)) exhibit the most elaborate treatment of the electronic structure problem and predict the experimentally measured dissociation energy of the hydrogen fluoride dimer to highest precision [51]. In addition, this observation demonstrates the methodological consistency of combining the (parameter-free) qce model with cluster results obtained from high level electronic structure calculations and indicates that in the case of hydrogen fluoride the estimation of the boiling point of the real liquid is possible on the basis of a pure ab initio treatment.

The phase transitions obtained from this parameter-free qce model as illustrated in Figs. 4.6 and 4.7 of course only qualitatively reflect the liquid phase behavior of the real liquid phases, but it is clear that even a partial consideration of interparticle interactions as accomplished in terms of the cluster structures is sufficient for a volume collapse to a low density phase, and that *no* volume for the particles constituting the system under investigation is necessary for this condensation phenomenon. It is furthermore apparent from Fig. 4.5 that even in the case of the most stable clusters (as obtained from the BP/TZVP combination) the phase transition temperature of real water is underestimated by at least 100 K, whereas the boiling point of real hydrogen fluoride is approximately captured by the high level quantum chemical cluster calculations. Following the reasoning pointed out before, this corresponds to an underestimation of the intermolecular interactions in the case of water, and it is apparent that in contrast to the hydrogen fluoride system the isolated cluster energetics as e.g. reflected in the values from Table 4.1 do not suffice to adequately model the extended intermolecular interaction network predicted to occur in the liquid water phase [57]. However, considering the fact that the zeroth order qce model explicitly neglects any possible interactions between the different clusters, this underestimation of the boiling temperature should be reducible by introducing additional stabilizing interactions between the cluster structures, i.e., by setting $a_{mf} > 0$ (J $\times$ m^3)/mol. Apparently this is not as necessary in the case of liquid hydrogen fluoride as for the qce treatment of water, which is possibly based on the more simple molecular structure of hydrogen fluoride exhibiting only a single donor functionality. The possibility of the water molecule to act as a double hydrogen bond donor is realized for almost none of the water molecules in the 7w8cube cluster set (see Fig. 4.1), and in the case of the spiro compounds only a single monomer unit per cluster structure exhibits the full fourfold coordination pattern. It is thus reasonable to expect that a larger fraction of intermolecular interactions is neglected in the case of the small to medium sized water clusters considered in this thesis as compared to the hydrogen fluoride cluster set, for which a twofold coordination pattern in the form of hydrogen-bonded chains is experimentally predicted [24, 25]. Even if recent experimental measurements demonstrate that the majority of water molecules assumes a twofold coordination in liquid water at ambient conditions as well, additional (non-hydrogen bond) interactions to the unoccupied donor and acceptor positions from the second solvation shell or in terms of long range interactions seem likely, and it is clear that such contributions are not considered in the isolated cluster approach applied in this section [10, 58]. Thus, the inclusion of residual cluster–cluster interactions can be expected to be of higher importance for an accurate qce treatment of liquid water, and the effect of the simple treatment of intercluster interactions according to Eq. 2.51 will be examined in Sect. 4.1.3.

4.1.2.2 Unmodified Excluded Volume

The behavior of an unscaled volume estimate ($a_{\mathrm{mf}} = 0$ (J $\times$ m^3)/mol, $b_{\mathrm{xv}} = 1$) in the frame of the qce model in combination with the 7w8cube cluster set has already been extensively examined in previous studies and therefore will not be reiterated in this thesis [15, 27, 59]. The extension of the 7w8cube set in terms of the spiro clusters again improves the convergency behavior in the low temperature domain, but does not otherwise affect the calculated isobars, leaving the computed molar volumes at a given temperature almost unchanged. The reason for this observation can be found in the relatively small values obtained for the unmodified excluded volumes according to Eq. 2.48, which are typically in the order of magnitude of 10^{-2} L and therefore do not alter the translational volume in the expression for the corresponding partition function (Eq. 2.49) significantly. The same situation is observed in the case of the unscaled excluded volume contributions calculated for the hydrogen fluoride cluster set, which are even smaller than the ones obtained for the extended water cluster set due to the smaller average number of monomer units per cluster. Thus, in the present setup in which the cluster volumes $\{V_j\}$ are estimated from the atomic van der Waals volumes (see the discussion of Eq. 2.48) and the cluster structures are comparably small, the isobars calculated in the frame of the parameter-free qce model are largely affected by the cluster interaction energies rather than the cluster volumes. However, this trend will possibly change if larger cluster structures are considered in the cluster set or if a more refined treatment for the determination of the cluster volumes with regard to structural issues is applied [29, 60].

4.1.3 Mean Field Attraction and Adjustment of Parameters

The results presented in the previous section demonstrate that the partial treatment of interparticle interaction in terms of isolated cluster structures (i.e., the ideal cluster gas) exhibits the qualitative features of a condensation to a phase of higher density as the one of a conventional ideal gas for both the examined substances water and hydrogen fluoride. Even if the calculated molar volumes and the condensation temperatures in most cases do not agree to the corresponding experimental values, these findings indicate that the cluster equilibrium approach is in principle capable of treating the condensed phase as well as the transition to the gas phase. The possibility of calculating phase transition properties in the frame of the qce model will be examined in greater detail in the following chapter. This section focusses on the effect of the mean field interaction between the clusters in order to increase the degree of accuracy of the qce isobars in the liquid phase temperature domain. For both examined substances, the high density phases found in the parameter-free qce approach are predicted to possess a molar volume of approximately 2 L in the low temperature domain regardless of the magnitude of the supramolecular cluster interaction energies (with the exception of the MP2/TZVPP pair interaction energies), see Figs. 4.6 and 4.7, which is approximately one order of magnitude smaller than the ideal gas volume. However, the molar volume of real water is still about two orders of magnitude smaller than the parameter-free qce result (approximately 1.8×10^{-2} L at room temperature), and this additional increase in density has to be accounted for via the mean field cluster interaction in the frame of the qce model [61]. Previous studies demonstrate that the molar volume of liquid water can be calculated to good precision on the basis of the 7w8cube cluster set if the mean field attraction parameter a_{mf} is adjusted manually to reproduce the experimental reference volume at a given temperature [15, 27, 59]. In the following, the effect of the mean field cluster interaction will be examined in a most general way and subsequently a more elaborate fitting procedure as outlined in Sect. 2.2.1 will be employed in order to obtain molar volumes of higher precision over larger temperature intervals for liquid water and liquid hydrogen fluoride [1, 50].

4.1.3.1 Unmodified Intercluster Interaction and Excluded Volume

Following the conception of the preceding part, the qualitative effect of the intercluster mean field interaction can be studied from a general perspective by considering an unmodified qce setup in which both the excluded volume contribution (Eq. 2.48) as well as the mean field interaction (Eq. 2.51) enter the calculation as unscaled quantities (i.e., setting $a_{\mathrm{mf}} = 1$ (J $\times$ m^3)/mol and $b_{\mathrm{xv}} = 1$). The isobars calculated for this parameter setup and the 7w8cube cluster set are

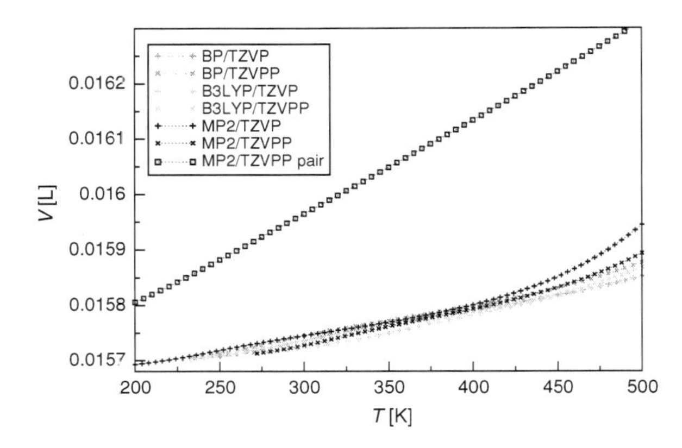

Fig. 4.8 Molar volume V as a function of the temperature T as obtained from different quantum chemical methods employed for the calculation of the 7w8cube cluster set. Excluded volume and intercluster interaction are unscaled ($a_{mf} = 1$ (J × m^3)/mol, $b_{xv} = 1$). The lowest temperature of each isobar corresponds to the point at which the qce iterations no longer converge (maximum number of iterations $n_{cyc} = 1{,}000$). All calculations refer to a pressure of $p = 1 \times 10^5$ Pa

illustrated in Fig. 4.8.[3] From these plots it is apparent that the inclusion of the unscaled mean field interaction leads to a shift of all calculated isobars to considerably smaller volumes which in all cases are even smaller than the molar volume of real liquid water. In addition, the indicated transition to the gas phase predicted for the case where the intercluster interaction is completely neglected (see Fig. 4.5) is no longer found in any of the curves even though the temperature interval for all qce calculations has been extended to $T = 500$ K. There is an increase in curvature visible in the isobar obtained from the MP2/TZVP cluster calculations at temperatures larger than $T = 450$ K, but not to such an extent as predicted for the complete neglect of the intercluster interaction. All curves obtained from the fully cooperative cluster calculations show a similar progression over the investigated temperature interval, with the MP2 isobars predicting the largest volumes at elevated temperatures. Smallest volumes are found for the MP2/TZVPP isobar and the B3LYP/TZVP isobar at lower temperatures and for the BP/TZVP as well as the B3LYP/TZVP curves at larger temperatures. The lowest temperatures at which the qce iterations still converge for a given quantum chemical cluster treatment (i.e., the low temperature ending points of the isobars) differ considerably between the employed methods. The largest stability with regard to this aspect is found for the MP2/TZVP combination, whereas the

[3] Please note that the isobars obtained from the extended water cluster set (7w8cube + spiro) qualitatively agree to the ones depicted in Fig. 4.8 and in most cases again show a better convergency behavior in the low temperature interval, but due to increased convergency problems for the MP2/TZVPP clusters at several temperatures the isobars of the original 7w8cube set are discussed here.

B3LYP/TZVP 7w8cube cluster set no longer converges at temperatures smaller than $T = 286$ K, which is close to the freezing point of real water.

However, the same trend can be observed in the case of the complete parameter-free setup (see Fig. 4.5) and therefore cannot be attributed to the intercluster interaction. The isobar calculated on the basis of the MP2/TZVPP pair interaction energies again differs from the remaining curves to a large extent. As in the case of the neglected intercluster interaction, the pair interaction energies lead to a lower density than the ones predicted for the fully cooperative interaction energies at all temperatures, with the differences being considerably larger than those found between the different method/basis set combinations. Furthermore, the pair interaction curve rises more steeply than any of the other isobars and again shows the ideal gas-like linear progression already observed for the pair interaction energies in the previously investigated parameter setup. This indicates that even though the pair interaction isobar in absolute terms is closest to the experimental one, a too linear temperature dependancy of the volume is obtained from these energies which does not exhibit the characteristic curvature found for many of the remaining isobars and which is also present in the isobar of real liquid water [62]. The comparison of these observations to the magnitudes of the interaction energies listed in Table 4.1 shows that in contrast to the parameter-free setup (for which the *temperature* of the condensation point is found to be dependent on the stability of the clusters), the degree of intracluster interaction now affects the *density* of the calculated phases. This is clearly seen in the high temperature domain, in which the methods yielding the most stable cluster structures (BP/TZVP and B3LYP/TZVP) also lead to the isobars of smallest molar volume and vice versa. At lower temperatures this trend is no longer as pronounced as in the high temperature domain, but the MP2/TZVP curve still exhibits the largest volume of all isobars based on the cooperative energies and the smallest volume is always found for the B3LYP/TZVP energies, which along with the BP/TZVP combination predict the most stable cluster structures. The considerable smaller magnitude of the pair interaction energies as well as the large volume difference between the corresponding pair interaction isobar and all remaining curves additionally support this observation, which has also been found in a similar form in previous qce calculations [27].

From these exemplary results it can thus be inferred that the stability of the cluster phase as expressed in the supramolecular interaction energies of the different clusters as well as in the mean field attraction largely determines the qualitative liquid phase behavior on both the temperature and the volume axis, and that the contribution of the excluded volume affects this behavior to a much lesser extent. With regard to this point it should be noted that the isobars obtained from the different methods not only depend on the calculated supramolecular cluster interaction energies, but also on the optimized structures and the harmonic frequencies which are both method-dependent and which enter the cluster partition functions as pointed out in Sect. 2.1.2. However, the example of the pair interaction energies and the corresponding isobars demonstrates that in the frame of the qce model the interaction energies apparently have a considerable impact on the

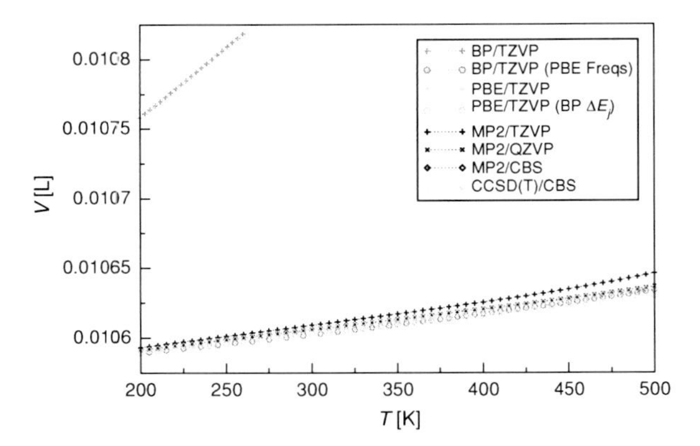

Fig. 4.9 Molar volume V as a function of the temperature T as obtained from different quantum chemical methods employed for the calculation of the hydrogen fluoride cluster set (see Fig. 4.3). Excluded volume and intercluster interaction are unscaled ($a_{\mathrm{mf}} = 1$ (J × m^3)/mol, $b_{\mathrm{xv}} = 1$). The lowest/highest temperature of each isobar corresponds to the point at which the qce iterations no longer converge (maximum number of iterations $n_{\mathrm{cyc}} = 1{,}000$). All calculations refer to a pressure of $p = 1 \times 10^5$ Pa

isobaric behavior, since the pair interaction curves are based on the same structures and frequencies as the fully cooperative MP2/TZVPP curves, but significantly differ from these as can be seen e.g. in Figs. 4.5, 4.6, and 4.8.

The isobars calculated for the hydrogen fluoride cluster set within the unmodified intercluster interaction setup ($a_{\mathrm{mf}} = 1$ (J × m^3)/mol, $b_{\mathrm{xv}} = 1$)) are illustrated in Fig. 4.9. As in the case of the water qce calculations, the activation of the mean field intercluster interaction leads to molar volumes approximately two orders of magnitudes smaller than the ones obtained for the isolated cluster gas. Furthermore, it is apparent that the behavior of the isobar obtained from the BP/TZVP cluster calculations is in strong contrast to the curves predicted by the remaining methods insofar as the calculated volume is larger by approximately 0.1 mL and the isobar rises more steeply than any of the remaining curves. At temperatures larger than $T = 260$ K the qce iterations no longer converge with the BP/TZVP cluster data as input, which is a surprising result considering the fact that the qce partition functions rely on the high temperature limit and the qce model in general shows a much improved convergency performance in the high temperature domain. There is no straightforward correlation visible between the BP/TZVP cluster interaction energies in Table 4.3 and the outlying behavior of the corresponding isobar, because all computed BP/TZVP cluster interaction energies are intermediate between the PBE/TZVP values and the numbers obtained from the MP2 calculations. Instead, the reason can be traced back to the harmonic frequencies obtained from the BP/TZVP normal mode analysis as can be seen from the isobar labeled as "BP/TZVP (PBE Freqs)" in Fig. 4.9. For this particular curve the cluster interaction energies (and principal moments of inertia) as obtained from the BP/TZVP calculations have been employed in combination with the frequencies calculated from the PBE/TZVP approach, and from this numerical experiment it is apparent that the convergency performance is improved and that the predicted volumes are considerably closer to the values obtained from the remaining methods. Thus, the isobar calculated from the BP/TZVP cluster data is very sensitive to the magnitude of the frequencies, and it is clear that not only the interaction energies, but also the set of harmonic frequencies can influence the qce isobars to a major extent. This is also seen in the isobar calculated from a modified approach (PBE/TZVP structures and frequencies combined with the BP/TZVP interaction energies, see the curve labeled as "PBE/TZVP (BP ΔE_j)"), which differs from the "BP/TZVP (PBE Freqs)"-isobar in the principal moments of inertia, but nevertheless shows a high agreement to this curve over a large temperature interval. In addition, if only the principal moments of inertia are taken from the PBE/TZVP calculations and the BP/TZVP frequencies and interaction energies are unmodified (not shown), the obtained isobar is identical to the original BP/TZVP curve showing the same overestimation in volume with regard to the remaining isobars and also an equally problematic convergency behavior. These artificial modifications thereby show that in the case of the hydrogen fluoride cluster set the changes in principal moments of inertia do not alter the calculated isobars to a large extent and that the effect of the interaction energies and harmonic frequencies on the molar volume is much more pronounced. With the exception of the outlying behavior of the BP/TZVP curve all remaining isobars show a continuous progression over the investigated temperature range at considerably lower volumes than for a complete neglect of the intercluster interaction, see Fig. 4.7. As in the case of the water isobars plotted in Fig. 4.8, the method predicting the most unstable cluster structures (MP2/TZVP) also yields the largest molar volumes, whereas all remaining curves only show very small differences over the whole temperature range. This observation is in contrast to the results obtained for the 7w8cube cluster set, where differences in the supramolecular cluster interaction energies in the order of approximately 10 kJ/mol and larger lead to visible differences between the corresponding isobars, see Table 4.1 and Fig. 4.8. In the case of the hydrogen fluoride cluster set even larger differences in the cluster interaction energies (approximately 60 kJ/mol and larger for the **c8r** cluster and the MP2/QZVP and PBE/TZVP methods, see Table 4.3) do not result in significantly different isobars as can be seen from Fig. 4.9. This result again indicates that for hydrogen fluoride the magnitude of the supramolecular cluster interaction energies does not have such a large impact as the one found in the case of the 7w8cube water cluster set and that in this case the influence of the harmonic frequencies is of increased importance.

The qce isobars obtained from the unscaled mean field intercluster interaction for both hydrogen fluoride and water predict molar volumes which are lower than the corresponding values of the real liquids for almost all methods at all considered temperatures, i.e., the density of the calculated phases is overestimated with respect to the experiment. According to the observations presented in this section so far, this indicates that the degree of attractive interactions within the van der Waals-like cluster gas is too large, which can either be a result of an overestimation of the cluster stability as expressed in the supramolecular cluster interaction energies or an overestimation of the mean field intercluster interaction. Given that the latter energy contribution is explicitly formulated in terms of a scaling factor (a_{mf}) in the qce approach, the performance of the method in reproducing the experimental densities via the adjustment of the a_{mf} scaling parameter will be examined in the following part.

4.1.3.2 Adjustment of the Model Parameters

The results presented so far demonstrate that it is possible to model the condensed phase behavior of associated liquids on the basis of the qce iterations in a qualitative way. In order to evaluate the capabilities of this approach for the calculation of real liquid phase properties the two scaling parameters a_{mf} and b_{xv} will be adjusted in the following according to the fitting procedure introduced in Sect. 2.2.1. As pointed out there, experimental isobars will be employed as reference curves for the determination of a set of error vectors $\{\mathbf{V}_k\}$ and the corresponding Euclidean norm $||\Delta\mathbf{V}||_k$ (the accuracy of the isobar k) over a fixed grid of a_{mf}-/ b_{xv}-values, and the isobars exhibiting the smallest $||\Delta\mathbf{V}||$ values are identified as the most accurate results with respect to the experimental reference curve [1]. The details of this adjustment procedure (parameter interval and stepsizes) are summarized in Sect. 7.1.

The isobars calculated in this way for the 7w8cube water set are illustrated in Fig. 4.10 along with the experimental reference isobar employed for the parameter adjustment [61]. From these graphs it is apparent that all quantum chemical methods employed for the calculation of the cluster properties are capable of reproducing the density of real liquid water over a large temperature interval in an accurate way, even though there are significant differences between the various methods in different temperature domains. The largest deviations from the experimental curve are found in the low temperature part near the freezing point for all employed methods, which predict a monotonous decline of the molar

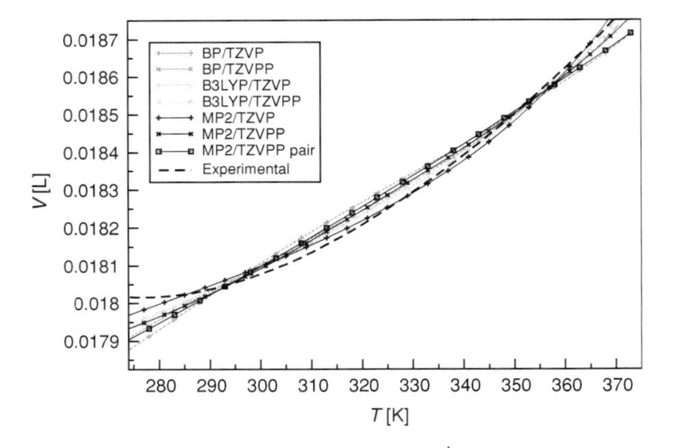

Fig. 4.10 Molar volume V as a function of the temperature T as obtained from different quantum chemical methods employed for the calculation of the 7w8cube cluster set compared to the experimental reference isobar [61]. The qce parameters are fitted to reproduce the experimental volume (see Table 4.4). All calculations refer to a pressure of $p = 1 \times 10^5$ Pa

volume without exception. The flattening of the experimental isobar as well as the point of maximum density is therefore not reproduced by any of the applied methods, but as stated previously the fourfold coordination pattern is an important motif in this temperature domain, and this bonding situation is not present in the 7w8cube cluster set [3, 16].

A closer inspection of the low temperature section shows that the curves based on the BP density functional yield the largest deviations from the experimental reference, whereas the isobars obtained from the MP2/TZVP and B3LYP/TZVP cluster calculations are the most accurate ones. However, a simple correlation between this observation and the magnitudes of the supramolecular interaction energies from Table 4.1 can no longer be drawn, because these energies are augmented with different mean field intercluster interaction contributions according to Eq. 2.52 as expressed in the different values of the mean field parameter a_{mf} listed in Table 4.4. This effect is also seen in the case of the MP2/TZVPP pair curve, which exhibits a similar progression as the BP/TZVP isobar over the whole temperature interval even though the corresponding cluster interaction energies show very large differences as can be seen from Table 4.1. In the mid-temperature interval all calculated isobars cross the experimental curve at approximately $T = 290$ K and after this point predict volumes too large with regard to the reference curve, thereby failing to reproduce the pronounced curvature of the experimental isobar. With respect to this point the BP curves as well as the MP2/TZVPP pair isobar again show the largest deviations.

There is another crossing region between the experimental reference and the calculated isobars in the high temperature domain starting at approximately $T = 330$ K (MP2/TZVP) and ranging up to $T = 350$ K (B3LYP/TZVP, MP2/TZVPP pair), after which all qce isobars again underestimate the experimental molar volume. However, at these temperatures the agreement between calculated

Table 4.4 Optimized qce parameters a_{mf} and b_{xv} as well as the accuracies $||\Delta \mathbf{V}||$ obtained from the adjustment procedure for both water cluster sets introduced in Sect. 4.1.1

Method	$\|\|\Delta \mathbf{V}\|\|$	a_{mf}	b_{xv}
7w8cube			
BP/TZVP	0.53	0.1207	1.084
BP/TZVPP	0.34	0.1308	1.084
B3LYP/TZVP	0.26	0.1338	1.101
B3LYP/TZVPP	0.25	0.1399	1.095
MP2/TZVP	0.18	0.1620	1.099
MP2/TZVPP	0.31	0.1620	1.107
MP2/TZVPP pair	0.44	0.3599	1.075
7w8cube + spiro			
BP/TZVP	0.57	0.1069	1.102
BP/TZVPP	0.40	0.1072	1.105
B3LYP/TZVP	1.18	0.1079	1.112
B3LYP/TZVPP	0.38	0.1070	1.103
MP2/TZVP	0.19	0.1550	1.117
MP2/TZVPP	0.14	0.1324	1.106
MP2/TZVPP pair	0.44	0.3599	1.075

$||\Delta \mathbf{V}||$ in [mL] and a_{mf} in [(J × m^3)/mol]

and experimental values is considerably higher as compared to the low temperature region, and the isobars obtained from the MP2/TZVP and B3LYP/TZVPP cluster calculations only show very small deviations in the order of magnitude of approximately 0.01 mL. In contrast, the BP/TZVP and MP2/TZVPP pair curves again show the largest differences, and considering the whole liquid phase temperature interval these isobars are the ones which underestimate the curvature of the experimental isobar in a most pronounced way. This behavior is also clearly reflected in the different magnitudes of the accuracies as obtained from the parameter adjustment procedure, see Table 4.4. From the first block it is apparent that the largest deviations (as expressed in large $\|\Delta\mathbf{V}\|$ values) are calculated for the BP/TZVP and the MP2/TZVPP pair isobar. In addition, the most accurate curve ($\|\Delta\mathbf{V}\| = 0.18\,\mathrm{mL}$) for the 7w8cube cluster set is obtained from the MP2/TZVP cluster calculations, which is clearly visible in Fig. 4.10. In the case of the dft methods, the enlargement of the basis set for the cluster calculations always results in more accurate qce isobars, whereas larger deviations in the curves are obtained by increasing the basis set from TZVP to TZVPP in the case of the MP2 approach. This is not necessarily an indication of methodological inconsistencies, but rather points at possible deficiencies in the cluster set itself as can be seen in the corresponding MP2/TZVP and MP2/TZVPP entries in the second block of Table 4.4. The comparison between the dft approaches and MP2 shows that the results obtained from B3LYP and MP2 are of comparable precision, whereas the calculated BP isobars are less accurate. Considering the significant differences observed for the MP2/TZVPP pair isobars in the previous parameter setups (see Figs. 4.6 and 4.8), the accurate performance of this approach in comparison to the fully cooperative calculations is worth mentioning. With regard to this point it is apparent that the methods predicting the least stable cluster structures (MP2/TZVP and MP2/TZVPP pair) exhibit the largest mean field parameter values (a_{mf}), and that the smallest values are found for the BP/TZVP combination, from which the most stable clusters are obtained. Thus, a possible underestimation of the intermolecular interaction on the *intra*cluster scale is compensated by an increased contribution on the *inter*cluster level via the adjustment procedure, thereby providing a way to account for the neglected cooperativity in the case of the MP2/TZVPP pair interaction energies. The resulting $\|\Delta\mathbf{V}\|$ value calculated by the adjustment procedure is comparable to the values obtained for the fully cooperative methods, which indicates that the missing cooperativity can be compensated in terms of the mean field approach in an efficient way, at least concerning the predicted molar volumes. In general, the a_{mf} values obtained from the adjustment procedure for the 7w8cube cluster set all lie in a narrow range between $a_{\mathrm{mf}} = 0.12$ (J $\times$ m^3)/mol and $a_{\mathrm{mf}} = 0.17$ (J $\times$ m^3)/mol (with the exception of the MP2/TZVPP pair isobar) and are thereby smaller than in the unmodified intercluster interaction setup ($a_{\mathrm{mf}} = 1$ (J $\times$ m^3)/mol) by nearly one order of magnitude. For this latter setup a considerable underestimation of the real liquid phase volume of water is observed as well as a decline of the calculated molar volumes with increasing magnitudes of the supramolecular cluster interaction energies, see Fig. 4.8. The comparison of these observations to the results obtained from the adjustment procedure in Table 4.4 indicates that this relation is also given in the case of the mean field intercluster interaction, i.e., smaller a_{mf} values yield larger molar volumes if the energy contribution from the intracluster level is kept constant. The adjusted excluded volume parameters b_{xv} all lie in an even more narrow range than the a_{mf} values between $b_{xv} = 1.07$ and $b_{xv} = 1.11$, which is relatively close to the unscaled volume estimate setup ($b_{xv} = 1$). This observation again emphasizes the previously stated result that in the case of water the proper treatment of the interparticle interactions is most important for modeling the liquid phase behavior, and that the consideration of the molecular volume affects the calculated isobars to a lesser extent.

The extension of the 7w8cube water cluster set in terms of the spiro clusters (see Fig. 4.2) shows a considerable effect on the results obtained from the adjustment procedure, see the second block in Table 4.4. The isobars calculated for this extended cluster set are illustrated in Fig. 4.11. These isobars indicate that the distinct curvature of the experimental isobar is reproduced in a more pronounced way for most of the applied methods in combination with the extended water cluster set, which is especially apparent in the low temperature domain where most calculated curves are closer to the experimental reference. In this temperature region the isobars based on the fully cooperative MP2 calculations as well as the B3LYP/TZVPP curve reproduce the experimental volume to highest precision, followed by the results obtained from the BP cluster calculations. Larger discrepancies can be found for the B3LYP/TZVP curve as well as the MP2/TZVPP pair curve, which again predicts an almost linear increase of the molar volume with temperature.

In comparison to the curve calculated for the 7w8cube cluster set, the isobar based on the B3LYP/TZVP cluster calculations is much less accurate in the case of the extended set over the whole temperature range, which demonstrates that the consideration of additional cluster structures can result in a decrease of the performance with respect to the experimental reference in certain cases. This behavior has been observed in the frame of the qce model before, and the generation of an "optimal" cluster set from a given variety of clusters via the elimination of underpopulated cluster structures can help to overcome this problem [2, 3]. In the

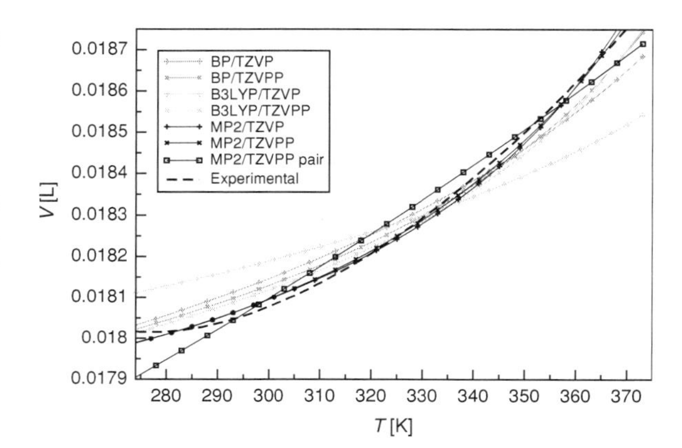

Fig. 4.11 Molar volume V as a function of the temperature T as obtained from different quantum chemical methods employed for the calculation of the extended (7w8cube + spiro) cluster set compared to the experimental reference isobar [61]. The qce parameters are fitted to reproduce the experimental volume (see Table 4.4). All calculations refer to a pressure of $p = 1 \times 10^5$ Pa

mid-temperature domain almost all calculated qce isobars reproduce the experimental volume to a reasonable precision, with the B3LYP/TZVP and the MP2/TZVPP pair curves again deviating to the largest extent. At higher temperatures close to the boiling point, all dft based isobars again show larger discrepancies with respect to the experiment, whereas the curves based on the fully cooperative MP2 calculations maintain the accuracy of the mid-temperature part of the isobars. In addition, the differences found between the MP2/TZVP and MP2/TZVPP isobars in the case of the smaller 7w8cube cluster set (see Fig. 4.10) are no longer present in the case of the extended set in Fig. 4.11. Both curves only show very small differences near the boiling point and are virtually identical in the low temperature domain. The origin of this behavior could well be coincidental, but might also be based on a systematic convergence with regard to the composition of the cluster set and the calculated cluster properties. Additional systematic studies in this direction including an even larger cluster set would be helpful to elucidate this matter. The qualitative trends discussed so far are again quantified in the values obtained from the adjustment procedure for the extended water cluster set, see the second block of Table 4.4. It is apparent that the large discrepancies in the high and low temperature part observed for the B3LYP/TZVP isobar in Fig. 4.11 is reflected in an outstanding large $||\Delta\mathbf{V}||$ value of $||\Delta\mathbf{V}|| = 1.18\,\text{mL}$, which is the largest value found in all qce calculations of liquid water in this thesis. The obtained a_{mf} and b_{xv} values assume equal magnitudes as in the case of the remaining methods and the supramolecular interaction energies are also close to the ones predicted by the BP/TZVP combination (see Table 4.2), which indicates that the B3LYP/TZVP harmonic frequencies are the crucial reason for the outlying behavior of the corresponding isobar. This assumption can again be verified in terms of the artificial mixing of the B3LYP/TZVP interaction energies and the BP/TZVP frequencies, which leads to a more accurate isobar ($||\Delta\mathbf{V}|| = 0.50\,\text{mL}$, not shown) than the one obtained from the pure B3LYP/TZVP data. For the remaining isobars a heterogeneous trend in the obtained qce parameters can be observed with regard to the ones found for the smaller 7w8cube cluster set. All curves based on dft cluster calculations show a decrease in precision (larger $||\Delta\mathbf{V}||$ values) regardless of the applied basis set, even though the loss of accuracy is not as pronounced as in the case of the B3LYP/TZVP result. In contrast, the MP2 isobars roughly retain their precision in the case of the TZVP basis set and show an improved performance in the case of the TZVPP basis set. Accordingly, the accuracy found for the MP2/TZVPP isobar as calculated from the extended water cluster set ($||\Delta\mathbf{V}|| = 0.14\,\text{mL}$) is the most precise result of all method/cluster set combinations examined in this thesis and thereby corresponds to the fact that this combination is the most sophisticated one from the quantum chemical perspective. Furthermore, the values listed in Table 4.4 show that the isobar calculated for the MP2/TZVPP pair interaction energies is not affected by the extension of the cluster set in any way, which is also reflected in vanishing populations obtained for these cluster structures in the MP2/TZVPP pair qce calculations at all considered temperatures (not shown). A reason for the neglect of the spiro structures in the MP2/TZVPP pair setup could lie in the decreased stabilization of the individual

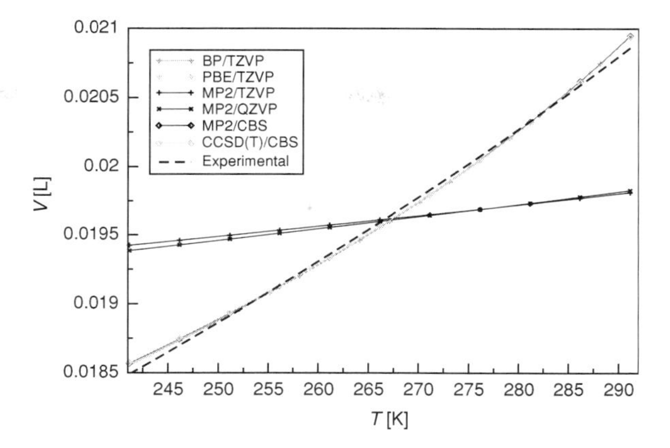

Fig. 4.12 Molar volume V as a function of the temperature T as obtained from different quantum chemical methods employed for the calculation of the hydrogen fluoride cluster set (see Fig. 4.3) compared to the experimental reference isobar [63]. The qce parameters are fitted to reproduce the experimental volume (see Table 4.5). All calculations refer to a pressure of $p = 1 \times 10^5$ Pa

water monomer expressed in the small magnitudes of the $\Delta E_{j,\mathrm{intra}}/i_j$ values (see the second block of Table 4.2) as compared to the remaining clusters, which is not present if the fully cooperative interaction energies are considered. The comparison of the different qce parameters found for the 7w8cube cluster set and the extended water cluster set indicates that with the exception of the MP2/TZVPP pair and the MP2/TZVP calculations the a_{mf} values are smaller by approximately 0.02–0.03 (J $\times$ m^3)/mol, whereas the b_{xv} values are larger by a similar magnitude in the case of the extended cluster set. The decrease of the intercluster mean field interaction energy can be rationalized in terms of the fourfold coordination pattern introduced in the spiro structures, which transfers a larger fraction of the intermolecular interaction to the intracluster level treated in terms of the quantum chemical calculations and thereby reduces the magnitude of the mean field intercluster interaction [2, 3]. In addition, the increase of the excluded volume scaling factor b_{xv} could possibly be an artifact of the rough van der Waals estimate for the cluster volume, which is expected to be more inaccurate in the case of expanded structures like the spiro clusters as compared to the rather compact structures of the 7w8cube set. A more refined treatment of molecular volumes could help to clarify this matter.

The isobars obtained from the parameter adjustment procedure for the hydrogen fluoride cluster set are illustrated in Fig. 4.12 together with the experimental curve [63]. It is immediately apparent that the cluster data obtained from the dft methods as well as the MP2/CCSD(T) cbs limit extrapolation calculations yield accurate isobars over the whole examined temperature interval, whereas the MP2 results obtained from the finite basis sets show considerable deviations to the experimental reference at most temperatures.[4] This behavior is yet another example for

[4] The temperature interval chosen for the hydrogen fluoride qce calculations only covers the upper half of the temperature range in which hydrogen fluoride is in its liquid state at $p = 1 \times 10^5$ Pa due to convergence problems of most applied methods at lower temperatures. A larger cluster set could help to improve these deficiencies as in the case of the extended water cluster set, see Figs. 4.5 and 4.6.

Table 4.5 Optimized qce parameters a_{mf} and b_{xv} as well as the accuracies $\|\Delta \mathbf{V}\|$ obtained from the adjustment procedure for the hydrogen fluoride cluster set introduced in Sect. 4.1.1

Method	$\|\Delta \mathbf{V}\|$	a_{mf}	b_{xv}
BP/TZVP	0.26	0.0208	1.351
PBE/TZVPP	0.22	0.0253	1.363
MP2/TZVP	4.16	0.0866	1.705
MP2/QZVP	4.05	0.0739	1.686
MP2/cbs	0.26	0.0255	1.357
CCSD(T)/cbs	0.27	0.0256	1.359

$\|\Delta \mathbf{V}\|$ in [mL] and a_{mf} in [(J × m^3)/mol]

the importance of the supramolecular interaction energies, since the MP2/QZVP and MP2/cbs qce calculations employ almost identical structures and harmonic frequencies. The outlying behavior of the MP2/TZVP and MP2/QZVP isobars is comparable to the result obtained from the B3LYP/TZVP approach in combination with the extended water cluster set (see Fig. 4.11) insofar as the experimental reference is reproduced to a reasonable precision in the mid-temperature domain, but the curves show larger deviations at lower and higher temperatures, respectively. This indicates that the slope of the experimental curves cannot be reasonably modelled by these approaches if no care of the basis set size is taken and that the obtained qce results are inadequate to describe the liquid phase behavior of the corresponding substances in the investigated temperature interval. In contrast, the isobars obtained from the cbs limit extrapolations (as well as those from the remaining dft methods) reproduce the experimental curve to high accuracy at almost all considered temperatures. A small increase in the deviations can again be observed in the low and high temperature domain, but more pronounced inaccuracies at lower temperatures as in the case of the 7w8cube water cluster set (see Fig. 4.10) are absent, which can either be attributed to the missing anomalous behavior of liquid hydrogen fluoride in this part of the phase diagram as compared to liquid water or in the fact that only the upper half of the liquid phase temperature interval of hydrogen fluoride is examined.

In addition, there are virtually no differences between the isobars computed on the basis of the dft and the MP2/CCSD(T) cbs cluster data, which is in contrast to the variation in the qce isobars calculated for the water cluster sets, see Figs. 4.10 and 4.11.[5] This indifferent behavior is also reflected in the optimized parameters as well as in the almost equal accuracy listed in Table 4.5. According to these values, the MP2 finite basis set results are considerably less accurate than any calculated water isobar (see Table 4.4), whereas the remaining results are of comparable accuracy. It is apparent that the loss of accuracy observed for the MP2/TZVP and MP2/QZVP isobars is accompanied by an increase in both the mean field interaction parameter a_{mf} as well as in the excluded volume parameter b_{xv}. This

[5] Please note that the scale of the volume axis in Fig. 4.12 differs from the one in Figs. 4.10 and 4.11, but even if the scale is aligned accordingly there are only very small differences observable between the MP2/CCSD(T) isobars and the curves based on the dft calculations.

behavior can again be rationalized in terms of the supramolecular interaction energies from Table 4.3, which indicate that the finite basis set MP2 calculations predict the most unstable cluster structures.

Accordingly, this underestimation of the intermolecular interactions on the intracluster scale is compensated by a larger contribution from the mean field intercluster interaction, but from the large $\|\Delta \mathbf{V}\|$ value it is apparent that this process cannot be realized as efficiently as e.g. in the case of the qce calculations based on the MP2/TZVPP pair interaction energies, see Table 4.4. The reversed relation between the magnitudes of the supramolecular interaction energies and the mean field interaction parameter is not directly present for the remaining methods, i.e., the most stable cluster structures (obtained from PBE/TZVP) correspond to an isobar exhibiting a small but not the smallest a_{mf} value. Thus, the effect from the differences in harmonic frequencies are again visible, but not as pronounced as the one arising from the varying supramolecular interaction energies. The magnitudes of the calculated excluded volume parameters b_{xv} again do not show large variations with respect to the applied quantum chemical methods (with the exception of the finite basis set MP2 results). In addition, these values are again relatively close to the unmodified excluded volume case ($b_{xv} = 1$) though larger deviations from this idealized setup are observed than in the case of the water calculations, see Table 4.4. With regard to this point it is apparent that all optimized b_{xv} values are larger than one, thereby indicating that the van der Waals spheres volume approach systematically underestimates the cluster volume. The a_{mf} values optimized for the hydrogen fluoride cluster set are smaller than the ones found for e.g. the 7w8cube water cluster set by again nearly one order of magnitude and thereby differ considerably from the unmodified setup ($a_{mf} = 1$ (J $\times$ m^3)/mol). This result indicates that the fraction of intermolecular interaction neglected by the isolated cluster calculations is smaller for liquid hydrogen fluoride than in the case of water, which again can be rationalized in terms of the more simple molecular structure of hydrogen fluoride regarding the possible number of hydrogen bond contacts as discussed before, see Sect. 4.1.2.

The results of the present section demonstrate that medium-sized cluster sets in combination with the qce approach are capable of reproducing the molar volumes of typical associated (hydrogen-bonded) liquids either qualitatively if the qce parameters are fixed to unity or in a semi-quantitative way if the parameters are adjusted to experimental reference values. Compared to the results of qce calculations published previously, the proposed adjustment procedure introduced in Sect. 2.2.1 leads to a much higher agreement between the computed isobars and the density of the real liquids over large fractions of the liquid phase temperature interval [15, 27, 59]. The presented results show that the experimental density of liquid water can be reproduced to within 1–3% regardless of the quantum chemical method employed for the cluster calculations and the experimental density of hydrogen fluoride in the high temperature domain of the liquid phase to within 1–2% if care of possible basis set effects is taken. These basis set effects are only present in the post-Hartree–Fock qce results and indicate that the consideration of polarization functions is more important than the presence of additional functions for the treatment of the valence levels as can be seen from the similar hydrogen fluoride isobars obtained from the MP2/TZVP (triple-ζ basis set) and the MP2/QZVP (quadruple-ζ basis set) cluster data. A possible underestimation of the supramolecular interaction energies can be corrected to a certain degree by an appropriate adjustment of the mean field interaction parameter as can be seen in the case of the MP2/TZVPP pair isobars for liquid water, but this possibility is not always given as indicated by the large deviations obtained for the hydrogen fluoride finite basis set MP2 isobars. Furthermore, the calculation of high quality supramolecular interaction energies can generally be expected to yield more accurate qce results than a possible compensation in terms of the a_{mf} parameter, because these energies are cluster-specific, whereas the mean field interaction parameter is adjusted for the whole cluster set and cannot account for e.g. structure-specific aspects of certain clusters. In addition, the effect of the limited size of the employed cluster set has to be considered as well, but this contribution is difficult to quantify. The results obtained from the extended water cluster set indicate that all important structural motifs have to be present within the applied cluster structures in order to model the liquid phase behavior over the whole temperature interval, and shortcomings of this kind could also be the reason for the missing convergence of the qce iterations in the low temperature half of the liquid hydrogen fluoride phase. It is also clear that the applied methodology is only suitable for the reproduction of the molar volume and not for an ab initio-like prediction, since the experimental density has to be available for the parameter adjustment procedure. However, the obtained parameters could subsequently be employed for the calculation of other thermodynamic properties beside the molar volume, and the practicability of this approach will be examined in the next section using the liquid phase entropy as a target quantity.

4.2 Liquid Phase Entropies from the Quantum Cluster Equilibrium Model

The general procedure for the calculation of condensed phase thermodynamic data in the frame of the qce model is explained in detail in Sect. 2.2.2. In order to evaluate the practicability of the derived equations and to obtain a performance test for the qce approach in predicting the liquid phase entropies of real systems, a routine for the calculation of entropies according to Eq. 2.68 has been implemented into the Peacemaker code [1, 37]. In analogy to the outline of the previous section, the entropies calculated according to the parameter-free qce setup will be discussed first, and the adjusted parameters from Tables 4.4 and 4.5 will be employed subsequently in order to assess the potential of this methodology for the quantitative calculation of condensed phase entropies. The systems under investigation are again the liquid phases of water and hydrogen fluoride.

4.2.1 Entropies from the Zeroth Order Model

The complete parameter-free model ($a_{\text{mf}} = 0$ (J $\times$ m^3)/mol, $b_{\text{xv}} = 0$) will again be employed first to gain information about the effect on the entropy arising from the transition to the (ideal) cluster gas in contrast to the single molecule picture of the conventional rrho approach. The qualitative influence of the intercluster mean field interaction will be studied next in terms of the unmodified mean field setup ($a_{\text{mf}} = 1$ (J $\times$ m^3)/mol). The results of the previous section indicate that the intercluster interaction has a more pronounced effect on the calculated isobars than the excluded volume contribution, which rather modulates the calculated curves and does not change the qualitative progression. In contrast, the analysis presented in Chap. 3 clearly demonstrates that the free volume of translation is a crucial quantity if the conventional rrho approach is employed for the calculation of condensed phase entropies. Thus, it will be interesting to see whether this pronounced influence is also present within the frame of the qce approach.

4.2.1.1 Parameter-Free Setup and Comparison to Weighting Approaches

The results presented in the previous section demonstrate that the formation of a high density liquid-like phase and and a low density gas phase can be observed in the frame of the complete parameter-free model ($a_{\text{mf}} = 0$ (J $\times$ m^3)/mol, $b_{\text{xv}} = 0$) solely due to the partial treatment of intermolecular interactions on the intracluster level, see e.g. Figs. 4.6 and 4.7. This effect is also clearly visible in the entropies calculated for the extended (7w8cube + spiro) water cluster set, which are illustrated in Fig. 4.13.[6] From this plot it is apparent that the calculated entropies show an almost identical progression over the investigated temperature interval as the corresponding isobars in Fig. 4.6, and the same trends are observed for the different quantum chemical methods employed for the cluster calculations. A closer examination of the computed numbers reveals subtle differences though. The condensation from the gas phase to the liquid-like phase occurs within a more narrow temperature interval in most cases, i.e., the phase transition more closely resembles a true first order phase transition, whereas the calculated condensation temperatures agree to the ones found for the parameter-free qce isobars. In the liquid-like temperature region small differences in the entropies between the applied quantum chemical methods can be observed in Fig. 4.13, which are not present in the corresponding isobars, see Fig. 4.16. The largest entropies in this temperature domain are predicted by the MP2/TZVP and the B3LYP/TZVPP cluster data, followed closely by the curve based on the BP/TZVPP calculations.

---

[6] The entropy curves calculated for the exclusive application of the 7w8cube cluster set qualitatively agree to the curves obtained from the extended cluster set with the exception of a lesser convergence performance at lower temperatures, which has also been found for the isobars, see Figs. 4.5 and 4.6. Thus, only the results calculated for the extended set will be discussed here.

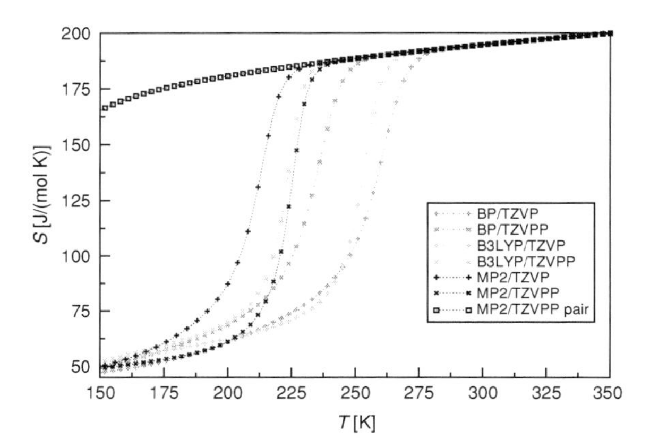

Fig. 4.13 Molar entropy S as a function of the temperature T as obtained from different quantum chemical methods employed for the calculation of the extended (7w8cube + spiro) cluster set. The lowest temperature of each curve corresponds to the point at which the qce iterations no longer converge (maximum number of iterations $n_{cyc} = 1{,}000$). All calculations refer to a pressure of $p = 1 \times 10^5$ Pa

In contrast, the BP/TZVP as well as the MP2/TZVPP combinations predict the smallest entropies in this low temperature region, which indicates that the observed differences neither arise from the different condensation temperatures predicted by the applied methods nor correlate with the varying cluster stabilities as expressed in the method-dependent supramolecular interaction energies in Table 4.1. The largest basis set effect with regard to these differences in the calculated entropies is found for the BP functional.

This effect is still visible at the lowest temperatures considered, whereas the entropies obtained from the remaining two methods no longer show relevant differences with regard to the applied basis set (TZVP or TZVPP) in this temperature region. However, these low temperature variations are not very pronounced with regard to the absolute magnitudes of the calculated entropies and should not be overemphasized. The comparison of the curves in Fig. 4.13 to the experimental entropy of liquid water ($S^0 = 70.0$ J/(mol K) at $T = 298.15$ K) reveals another difference between the molar volumes and the entropies obtained from the parameter-free qce setup [64]. According to the isobars plotted in Fig. 4.6, the liquid-like phase found for this setup overestimates the molar volume of real water by approximately two orders of magnitude regardless of the applied quantum chemical methodology. It is clearly seen from Fig. 4.13 that this considerable deviation is absent in the case of the entropies, which assume values between 50 and 60 J/(mol K) in the high density part of the phase diagram at temperatures below 200 K. It is apparent that in contrast to the behavior found for the molar volumes these numbers are in the same order of magnitude as the liquid phase entropy of real water. Thus, the parameter-free qce approach predicts a situation which is already close to the condensed phase behavior of real water in entropic terms, and the qualitative agreement indicates that the populations obtained from the underlying cluster equilibrium (see Eq. 2.54) in combination with the corresponding N-particle partition function (see Eq. 2.46) are able to mimic a real condensed phase behavior without any parameter adjustment. It is important to note that this result is only obtained from the N-particle partition

Table 4.6 Single cluster entropies for the extended (7w8cube + spiro) cluster set obtained from different methods and basis sets at $T = 298.15$ K according to the rrho model

Cluster	BP86		B3LYP		MP2	
	TZVP	TZVPP	TZVP	TZVPP	TZVP	TZVPP
w1	194.7	194.6	194.4	194.4	194.4	194.3
w2	283.7	285.2	278.5	272.1	288.6	288.9
w3A	320.3	322.1	320.6	322.5	327.9	329.4
w3B	324.2	324.2	326.5	327.7	332.4	332.7
w5	442.9	447.9	439.6	431.6	451.0	457.1
w6	504.4	507.6	501.6	511.2	519.1	528.6
s7	549.9	554.3	565.6	573.0	573.0	576.8
w8cube	521.8	532.6	550.0	544.0	544.9	547.7
s9	679.5	690.8	696.3	702.3	705.9	717.1
s11	818.7	855.1	835.9	842.3	843.5	861.9
s13	973.2	995.1	980.1	984.4	992.7	1019.2

All values are given in [J/(mol K)]

function and the corresponding entropy expression (see Eq. 2.68) which are based on the equilibrium between the different cluster structures and not from the isolated cluster calculations alone.

This is clearly seen if the entropies calculated from the cluster equilibrium approach are compared to the single cluster entropies calculated according to the conventional rrho approach, see Table 4.6 [1]. As pointed out in Sect. 2.1.2, the cluster partition functions employed in the calculation of these values are all based on the ideal gas volume and only differ in the cluster properties obtained from the quantum chemical calculation like e.g. the frequencies. Accordingly, the larger cluster structures also exhibit larger individual entropies as can be seen directly from the entropy contributions of the different cluster degrees of freedom as expressed in Eqs. 3.8, 3.9, and 3.10, i.e., a more expanded spatial structure will likely lead to larger principal moments of inertia and a cluster containing a larger number of atoms exhibits a higher number of normal modes and therefore a larger vibrational entropy. This behavior is clearly reflected in the values summarized in Table 4.6. The increase of the cluster size is directly reflected in a larger entropy value in all cases. The differences between the applied quantum chemical methods mainly arise due to differences in the vibrational entropy contributions and are most pronounced for the values obtained from the BP/TZVP and the MP2/TZVPP calculations, i.e., the latter method always predicts largest cluster entropies and BP/TZVP yields the smallest values for most of the clusters. Again basis set effects are clearly visible, and the largest ones are found for the BP and the MP2 approaches, whereas the differences in entropy obtained from B3LYP/TZVP and B3LYP/TZVPP do not exceed approximately 10 J/(mol K). The increase of the entropy with the cluster size amounts to approximately 60–80 J/(mol K) per water monomer for most of the structures belonging to the 7w8cube cluster set, but is considerably enlarged to more than 100 J/(mol K) in the case of the spiro structures, thereby showing the effect of the expanded geometry of these compounds.

A smaller increase in entropy is predicted for the transition from the dimer **w2** to the cyclic trimers **w3A** and **w3B**, which indicates that the ring closure taking place in this transition is entropically disfavored. Relating the numbers from Table 4.6 to the curves in Fig. 4.13 shows that none of the isolated cluster values can account for the low temperature/high density entropies obtained from the qce calculations, regardless which cluster population distribution is assumed. Even the exclusive population of the monomer exhibiting the smallest individual entropy would overestimate the parameter-free qce entropies (as well as the experimental value) by more than 100 J/(mol K) at the lowest considered temperature ($S_{\mathrm{w1}}(150\ \mathrm{K}) = 171.5$ J/(mol K)), and the results presented in Sect. 4.1.2 as well as previous studies on the subject demonstrate that the larger clusters are essential in the low temperature domain [2, 16]. In order to quantify this observation, a test calculation employing the cluster population distribution obtained from the parameter-free qce approach at $T = 200$ K as well as the individual cluster entropies calculated for $T = 200$ K on the basis of the MP2/TZVPP cluster data has been carried out, which yields a population-weighted entropy value of $S_{\mathrm{pop}}(200\ \mathrm{K}) = 398.8$ J/(mol K) for the extended water cluster set, whereas the corresponding MP2/TZVPP qce entropy is equal to $S_{\mathrm{qce0}}(200\ \mathrm{K}) =$ 61.1 J/(mol K). As previously assumed, this large overestimation is based on the considerable population of the large w8cube cluster which amounts to more than 75% and thereby also controls the entropy calculated according to this weighting approach. With regard to these findings it should be noted that the weighting of individual cluster properties in terms of population distributions obtained from qce calculations has been successfully applied before e.g. in the context of liquid phase quadrupole coupling parameters or hydrogen bond number studies [3, 26, 65]. However, in the case of condensed phase entropy calculations this approach is unsuitable as can clearly be seen from the presented numbers [1]. Considering the results from Sect. 4.1.2 as well as the chosen parameter setup ($a_{\mathrm{mf}} = 0$ (J $\times$ m^3)/mol, $b_{\mathrm{xv}} = 0$), this observation can again be rationalized in terms of the ideal gas volume applied for the calculation of the translational contribution to the individual cluster entropies in Table 4.6. Since both qce parameters are fixed to zero, possible influences from the excluded volume part or the intercluster mean field interaction can be excluded. In addition, the transition from the single molecule to the partial treatment of intermolecular interactions in terms of cluster structures cannot be the reason for the observed discrepancies, because the individual entropies listed in Table 4.6 also include these interactions as they are obtained from the same cluster structures. Consequently, the only difference between the rrho approach and the qce model in the parameter-free setup is given by the volume employed for the translational partition function (see Eq. 2.49) as well as the supramolecular interaction energies employed for the electronic partition function (see Eq. 2.52). A detailed analysis of this latter contribution to the condensed phase entropies calculated on the basis of the 7w8cube cluster set demonstrates that the effect of the electronic partition function is completely negligible, see Table 4.3 in Ref. [1]. Thus, it is again the proper treatment of the volume available for translational motion which is the prerequisite for obtaining a

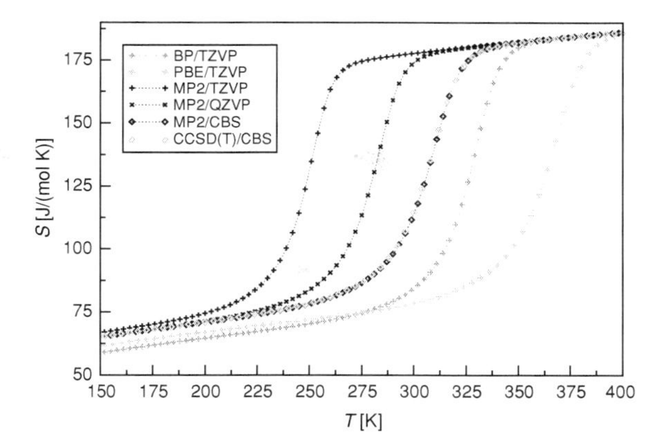

Fig. 4.14 Molar entropy S as a function of the temperature T as obtained from different quantum chemical methods employed for the calculation of the hydrogen fluoride cluster set (see Fig. 4.3). The lowest temperature of each curve corresponds to the point at which the qce iterations no longer converge (maximum number of iterations $n_{\mathrm{cyc}} = 1{,}000$). All calculations refer to a pressure of $p = 1 \times 10^5$ Pa

qualitative liquid phase behavior of the entropy. The application of the ideal gas volume in this regard again results in a considerable overestimation of the translational entropy as expressed in the values listed in Table 4.6 as well as in the population-weighted entropy of $S_{\mathrm{pop}}(200\ \mathrm{K}) = 398.8$ J/(mol K), which is in total agreement with the conclusions drawn in Chap. 3. With regard to this result it is interesting to see that the qualitative correct prediction of the liquid phase entropies according to the plots in Fig. 4.13 relies on the volume collapse found in Sect. 4.1.2 for the parameter-free qce setup, which in turn is a direct consequence of the partial treatment of intermolecular interactions within the cluster structures as well as the underlying cluster equilibrium. Thus, for the static modeling of condensed phase entropies it is apparently necessary to treat intermolecular interactions at least partly in order to obtain a reasonable volume for the system under investigation, from which the entropies can be obtained subsequently. If this is not possible for instance due to the system size (e.g. in the case of biomolecules or large supramolecular compounds) an appropriate scaling of the free translational volume as discussed in the context of the cell model approach in Sect. 3.2.3 could provide an alternative route. With regard to this point the comparison between the isobars and the entropies of the parameter-free qce model in Figs. 4.6 and 4.13 indicates that in order to obtain an order-of-magnitude estimate of the condensed phase entropy it is not necessary to employ a highly accurate value for the volume, since the isobars computed for the extended water cluster set in this setup still show considerable discrepancies to the density of real liquid water.

The entropies calculated for the hydrogen fluoride cluster set in the frame of the parameter-free qce setup are illustrated in Fig. 4.14. It is again apparent that the qualitative progression of the curves is very similar to the one observed for the corresponding isobars, see Fig. 4.7. As in the case of the entropies calculated for the parameter-free water model, the condensation to the high density liquid-like phase occurs within a smaller temperature interval. The sequence as well as the location of the condensation temperature predicted for the different methods exactly reproduces the situation found for the corresponding isobars, which is no surprise considering the pronounced effect of the volume on the calculated entropies as discussed above. The comparison of the curves in Fig. 4.14 to the entropies calculated for the extended water cluster set in Fig. 4.13 indicates that similar values of approximately 180 J/(mol K) are predicted for both substances in the gas phase. In contrast, the entropies obtained for the liquid-like hydrogen fluoride phase assume values between 60 and 65 J/(mol K), thereby being larger than the corresponding water results by approximately 15 J/(mol K). As in the case of the entropies obtained from the parameter-free water model, there are differences in the calculated low temperature entropies according to the quantum chemical methodologies applied for the cluster calculations, which are also present in the case of the corresponding isobars (see Fig. 4.7), but not to this extent. The curves based on the MP2 cluster calculations predict the largest entropies of approximately 65 J/(mol K) in this temperature domain regardless of the applied basis set, and the CCSD(T)/cbs curve also lies within this range. At the lowest temperatures considered in the qce calculations there are virtually no differences visible in the entropies obtained from the MP2/QZVP, MP2/cbs, and CCSD(T)/cbs cluster data, whereas the MP2/TZVP calculations predict a slightly larger value. In contrast, the entropies obtained from the dft cluster calculations assume constantly smaller values up to the phase transition temperature to the gas phase.

With regard to this observation it is apparent that the BP/TZVP combination yields the smallest entropies over a large fraction of the low temperature domain, but at the same time exhibits a lower phase transition temperature to the gas phase as compared to the curve predicted by the PBE/TZVP calculations, which results in an intersection of both curves at approximately $T = 275$ K. A similar behavior can be observed in the case of the entropies obtained from the parameter-free water model for the BP/TZVP and the B3LYP/TZVP curves, see Fig. 4.13. However, the approximately constant difference between the entropies based on the dft and MP2 calculations over a large part of the low temperature domain found in the case of the hydrogen fluoride results are not present in the entropies obtained for the extended water cluster set, which instead pass through several intersections between the MP2 and dft curves. As in the case of water, the results presented in Fig. 4.14 again indicate that the treatment of a condensed phase-like entropic behavior is principally possible if the equilibrium between the different cluster structures and the corresponding canonical partition function are considered, and the values obtained from the parameter-free hydrogen fluoride model are again in the same order of magnitude as the entropy of real liquid hydrogen fluoride [56]. The capability to calculate the liquid phase entropies of these two substances under investigation in an even more precise way via the optimized qce parameters from Sect. 4.1.3 will be examined in the following. However, a possible effect of considering the excluded volume contribution in an unmodified way ($b_{xv} = 1$) on the calculated entropies will be discussed first in the following paragraph, since the results presented so far clearly show that the entropies obtained from the rrho approach as well as from the qce model are very sensitive to this particular quantity.

4.2.1.2 Effect of the Unmodified Excluded Volume on the Entropy

The discussion presented in the final part of Sect. 4.1.2 indicates that the consideration of the unmodified van der Waals sphere cluster volume in combination with the neglect of the intercluster mean field interaction has virtually no effect on the calculated isobars, i.e., the molar volumes have been found to be equal to the complete parameter-free qce setup. The reason for that observation lies in the relatively small magnitude of the excluded volume itself, which is approximately two orders of magnitude smaller than the phase volume obtained from Eq. 2.63. With regard to these results, one could expect a minor effect of the unmodified excluded volume on the qce entropies at the most, and this reasoning is confirmed by qce calculations employing the corresponding parameter setup ($a_{mf} = 0$ (J $\times$ m^3)/mol, $b_{xv} = 1$) as well as the cluster sets introduced in Sect. 4.1.1. This behavior is exemplarily illustrated in Fig. 4.15 for the entropies obtained from the hydrogen fluoride cluster set. Comparing the calculated curves to the entropies predicted by the complete parameter-free approach ($a_{mf} = 0$ (J $\times$ m^3)/mol, $b_{xv} = 0$, see Fig. 4.14 and the grey curves in Fig. 4.15) clearly demonstrates that the unmodified excluded volume does not affect the obtained entropies in any way, and the same result is found for both water cluster sets. Thus, even though the results presented so far show a pronounced dependancy of the entropy on the volume in the condensed phase, variations in the order of magnitude of 1–5% as computed for the unmodified excluded volume contribution do not lead to significant variations in the final values. This finding could also be anticipated by considering the scaling of the inverse number density (i.e., the volume) in the translational entropy as expressed in Eq. 3.29 and Fig. 3.7. However, it should be

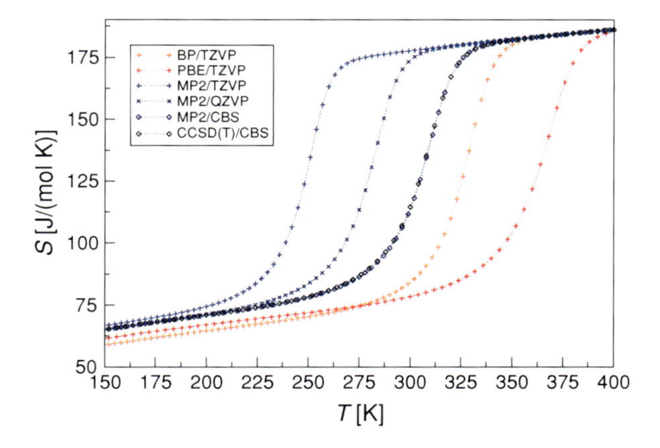

Fig. 4.15 Molar entropy S as a function of the temperature T as obtained from the unscaled excluded volume ($b_{xv} = 1$) and different quantum chemical methods employed for the calculation of the hydrogen fluoride cluster set (see Fig. 4.3). The lowest temperature of each curve corresponds to the point at which the qce iterations no longer converge (maximum number of iterations $n_{cyc} = 1{,}000$). The grey curves indicate the corresponding entropies from the parameter-free setup (see Fig. 4.14). All calculations refer to a pressure of $p = 1 \times 10^5$ Pa

noted that the qce entropy shows an additional dependancy on the volume introduced via the intercluster mean field interaction term (see Eqs. 2.67 and 2.68) in which the volume enters according to Eq. 2.51, but the parameter setup chosen for the present calculations eliminates this contribution due to the zero magnitude of the a_{mf} parameter.

4.2.2 Mean Field Attraction and Entropy

The results presented in the previous section show that in contrast to the molar volume discussed in Sect. 4.1.2 the experimental entropies of real liquid water and hydrogen fluoride are already predicted in a qualitative way by the complete parameter-free qce model. As in the previous part (Sect. 4.1.3) the focus is now set on the inspection of the potential to calculate these entropies quantitatively by considering the possibilities of scaling the excluded volume and the mean field intercluster interaction contributions. However, the effect of the unmodified intercluster interaction on the qce entropies will be examined first.

4.2.2.1 Effect of the Unmodified Correction Terms on the Entropy

The consideration of the unscaled intercluster interaction and the unscaled excluded volume ($a_{\mathrm{mf}} = 1$ (J $\times$ m^3)/mol, $b_{\mathrm{xv}} = 1$) has been found to affect the calculated molar volumes to a large extent in Sect. 4.1.3 in reducing the volume by approximately two orders of magnitude. From these results a pronounced effect on the qce entropies can be expected as well. Figure 4.16 illustrates the calculated entropies for the 7w8cube water cluster set. From these plots it is apparent that the obtained entropies are considerably affected by the mean field intercluster interaction though not to such an extent as observed in the case of the molar volumes in the first part of Sect. 4.1.3.

The most obvious difference to the curves predicted by the parameter-free setup in Fig. 4.13 is the absence of a phase transition to the gas phase, which is not found in the present case even if the temperature interval is further enlarged to $T = 800$ K (not shown). Instead, the calculated entropies show a homogeneous increase with temperature and converge to a high temperature limit of approximately $S = 140$ J/(mol K). A comparable behavior is also observed in the case of the molar volumes in Fig. 4.8, and these results can again be rationalized in terms of the additional energetic stabilization introduced via the (attractive) mean field cluster interaction as discussed previously. However, the large discrepancies in magnitude found between the complete parameter-free setup and the unmodified setup in the case of the isobars (see Figs. 4.5 and 4.8) is not present in the calculated entropies. As in the parameter-free case, the obtained entropy values of most methods lie within a narrow range between 60 and 80 J/(mol K) over a large fraction of the considered temperature interval. The only significant exception to

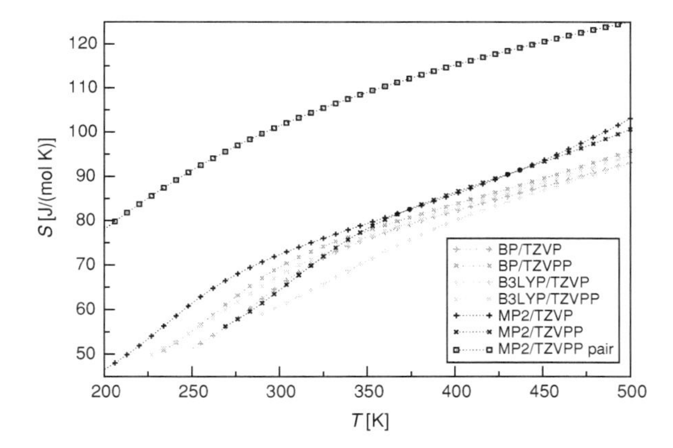

Fig. 4.16 Molar entropy S as a function of the temperature T as obtained from different quantum chemical methods employed for the calculation of the 7w8cube cluster set. Excluded volume and intercluster interaction are unscaled ($a_{\mathrm{mf}} = 1$ (J $\times$ m^3)/mol, $b_{\mathrm{xv}} = 1$). The lowest temperature of each curve corresponds to the point at which the qce iterations no longer converge (maximum number of iterations $n_{\mathrm{cyc}} = 1{,}000$). All calculations refer to a pressure of $p = 1 \times 10^5$ Pa

this observation is given by the entropies obtained from the MP2/TZVPP pair interaction energies, which qualitatively show a similar progression as the one of the remaining curves, but shifted to higher absolute values by a constant contribution of approximately 30 J/(mol K). The entropies calculated on the basis of the fully cooperative interaction energies again show method-dependent variations as already observed in the parameter-free setup, see Fig. 4.13. As before, the largest entropies are predicted by the MP2/TZVP cluster calculations over the whole temperature interval. At high temperatures, the values obtained from the MP2/TZVPP calculations are in close vicinity to the corresponding curve calculated from the MP2/TZVP cluster data, but at temperatures lower than approximately $T = 350$ K larger discrepancies are observed. This behavior is a general trend found in the present parameter setup, i.e., the differences between the calculated entropies obtained from the dft calculations increase at lower temperatures as well, whereas only small discrepancies are predicted in the high temperature domain. With regard to this observation it is seen that in the upper half of the considered temperature interval the basis set effects are larger than the methodological differences between the applied density functionals. This similarity in the entropies as obtained from the BP and the B3LYP cluster data is also present at lower temperatures in the case of the larger TZVPP basis set, whereas increasing discrepancies are observed between the two density functionals in the case of the TZVP basis set in this temperature domain. This difference is essentially based on the progression of the B3LYP/TZVP curve, which evolves nearly linearly and does not exhibit the pronounced curvature found for all remaining methods. This outlying behavior is additionally emphasized by the early convergence failure found for the B3LYP/TZVP qce iterations, which do not converge at temperatures lower than $T = 286$ K, see also Fig. 4.8. In contrast, the entropies calculated from the remaining cluster data show a progression increasing fast in the low temperature domain and exhibiting a characteristic change in the slope over a small temperature interval, which occurs in the mid-temperature domain approximately between $T = 275$ K and $T = 350$ K depending on the particular method and basis set combination and which is not present in the corresponding curves obtained from the parameter-free setup, see Fig. 4.13. This characteristic decrease of the slope can also be observed in the experimental entropy of real liquid water though to a lesser extent, and the results of the following section will show whether this qualitative agreement can further be improved via the application of the adjusted qce parameters from Table 4.4.

The entropies calculated for the hydrogen fluoride cluster set in the frame of the unscaled intercluster interaction and the unscaled excluded volume ($a_{\mathrm{mf}} = 1$ (J $\times$ m^3)/mol, $b_{\mathrm{xv}} = 1$) qce setup are plotted in Fig. 4.17. As in the case of the entropies obtained for the 7w8cube cluster set, largest values over the whole temperature range are found for the MP2/TZVP cluster data. There are virtually no differences between the curves obtained from the MP2 calculations employing the larger basis sets (QZVP, cbs) as well as from the CCSD(T) interaction energies, which is no surprise considering that also no differences are observed in the corresponding isobars, see Fig. 4.9. The progression of the entropy curves based on these cluster data is similar to the one found for the MP2/TZVP combination, but shifted to lower values by approximately 2.5 J/(mol K). Larger differences are predicted for the two dft approaches considered for the cluster calculations.

As in the case of the molar volume, the BP/TZVP combination yields considerable larger values than the ones obtained from PBE/TZVP, but in the present

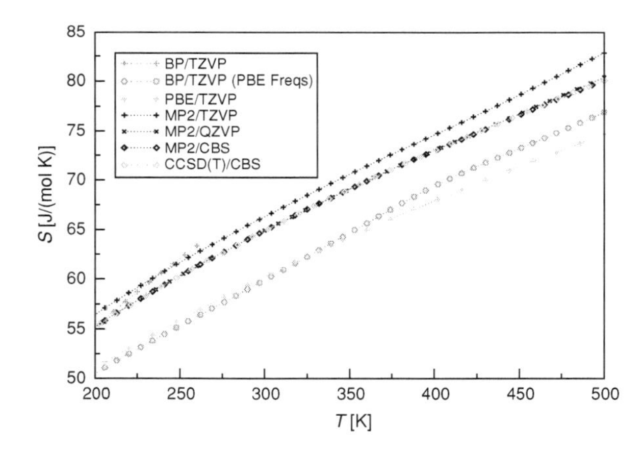

Fig. 4.17 Molar entropy S as a function of the temperature T as obtained from different quantum chemical methods employed for the calculation of the hydrogen fluoride cluster set (see Fig. 4.3). Excluded volume and intercluster interaction are unscaled ($a_{\mathrm{mf}} = 1$ (J $\times$ m^3)/mol, $b_{\mathrm{xv}} = 1$). The lowest/highest temperature of each curve corresponds to the point at which the qce iterations no longer converge (maximum number of iterations $n_{\mathrm{cyc}} = 1{,}000$). All calculations refer to a pressure of $p = 1 \times 10^5$ Pa

case the predicted values are similar to the MP2 and CCSD(T) curves and do not show the outlying characteristics found for the molar volume, see Fig. 4.9. If the improved convergence behavior obtained from the combination of the BP/TZVP interaction energies (and structures) with the PBE/TZVP harmonic frequencies is again employed for the calculation of the entropy, the discrepancies between the two dft approaches are significantly reduced in the low temperature domain and at higher temperatures exhibit a comparable magnitude as the ones between the MP2/TZVP curve and the remaining MP2 results. The entropy curve calculated on the basis of the PBE/TZVP cluster data shows a similar progression as the one of the MP2/cbs and the CCSD(T)/cbs curves, but is shifted to lower values by an essentially temperature-independent difference of approximately 5 J/(mol K). The comparison of the hydrogen fluoride entropies obtained from the unmodified qce setup to the corresponding results obtained in the frame of the complete parameter-free model (see Fig. 4.14) reveals similar changes as observed for the case of water. The transition to the low density phase is again absent, and the characteristic progression of the curves in Fig. 4.17 exhibiting a smaller slope at larger temperatures is also not reflected in the low temperature part of Fig. 4.14. However, with the exception of the BP/TZVP values, the sequence of decreasing entropies at a given temperature (MP2/TZVP → MP2/QZVP, MP2/cbs, CCSD(T)/cbs → PBE/TZVP) is predicted equally in both parameter setups. Comparing the curves plotted in Fig. 4.17 to the entropies calculated for the corresponding parameter setup and the 7w8cube cluster set (see Fig. 4.16), a qualitative agreement in the temperature dependancy of the entropies predicted for both substances is indicated, which is not observed in the case of the molar volumes to this extent, see Figs. 4.8 and 4.9. The magnitudes of the qce entropies assume values of approximately 50 J/(mol K) in the low temperature domain and rise to approximately 80 J/(mol K) over the examined temperature interval in both cases, with the MP2/TZVP cluster data predicting slightly larger entropies for water as well as for hydrogen fluoride. This is not necessarily to be expected, since the results presented in the previous parts of this chapter demonstrate non-negligible differences in the calculated molar volumes (see Figs. 4.8 and 4.9), in the supramolecular interaction energies (see Tables 4.1 and 4.3), and in the optimized qce parameters (see Tables 4.4 and 4.5) obtained for both substances. However, this observation again can be rationalized via the logarithmic dependancy of the entropy on the volume (see e.g. Eq. 2.66), from which a variation in entropy of approximately 4 J/(mol K) can be estimated according to the observed differences in the calculated molar volumes.

Thus, the consideration of the unmodified qce correction terms does not influence the calculated entropies to such a large extent as observed for the molar volumes in the previous section, and it is found that the ideal parameter setups considered so far already yield liquid phase entropies which are in the right order of magnitude with regard to the entropy of e.g. real liquid water. An additional improvement of these predictions can possibly be achieved via the application of the optimized parameters listed in Tables 4.4 and 4.5. This possibility will be examined in the following part.

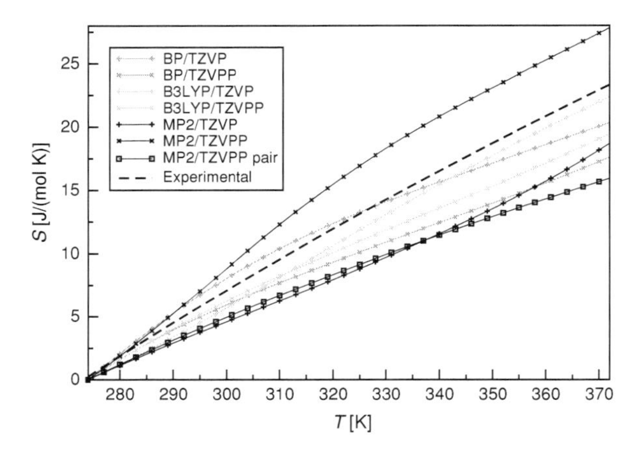

Fig. 4.18 Molar entropy S as a function of the temperature T as obtained from different quantum chemical methods employed for the calculation of the 7w8cube cluster set compared to the experimental entropies [61]. The reference for all curves is set to $S(273.15\ \text{K}) = 0$ J/(mol K). The qce parameters are fitted to reproduce the experimental volume (see Table 4.4). All calculations refer to a pressure of $p = 1 \times 10^5$ Pa

4.2.2.2 Entropies from the Adjusted Parameter Setup

The results obtained for the parameter adjustment with regard to the experimental density of liquid water indicate that the applied methodology is capable of reproducing the experimental reference to a reasonable precision, see the last part of Sect. 4.1.3. However, it is also important to assess the predictive capabilities of the obtained parameters in terms of a quantity which has not been applied in the parameter adjustment process. The liquid phase entropies of water and hydrogen fluoride will be employed for this task in the following. The comparison of the entropies calculated for the 7w8cube cluster set from the optimized parameters listed in Table 4.4 to the experimental entropy of liquid water is illustrated in Fig. 4.18. From these plots it is apparent that the experimental reference is predicted to a reasonable precision by all applied quantum chemical methods even though the average deviations are larger as in the case of the molar volumes, see Fig. 4.10. For almost all methods larger discrepancies are found in the high temperature part of the liquid phase interval. This observation constitutes a reversed trend as compared to the behavior obtained for the molar volume on the basis of the 7w8cube cluster set, where largest inaccuracies are predicted in the low temperature domain as can be seen from Fig. 4.10. The curves in Fig. 4.18 show that in this high temperature domain almost all applied methods underestimate the experimental reference and that the only overestimation of approximately 5 J/(mol K) is found for the MP2/TZVPP cluster data. The most accurate liquid phase entropies are obtained from the B3LYP/TZVP cluster calculations exhibiting differences of less than approximately 3 J/(mol K) over the whole liquid phase temperature range, which is in contrast to the intermediate performance of this combination with regard to the accuracies of the adjusted isobars, see Table 4.4.

A similar agreement can be observed for the entropies based on the BP/TZVP cluster calculations, which predict a too steep increase at low temperatures but underestimate the slope of the experimental curve at higher temperatures, thereby intersecting the experimental reference at approximately $T = 330$ K. Larger deviations are obtained from the dft methods in combination with the TZVPP basis set, which again is in contrast to the observation that the larger basis set yields more accurate isobars, see Table 4.4. With regard to this point it is also obvious that the method yielding the most accurate isobar for the 7w8cube cluster set (MP2/TZVP) shows largest discrepancies in the calculated entropies especially in the mid-temperature domain, which are only exceeded by the MP2/TZVPP pair and the BP/TZVPP curve at larger temperatures due to a sudden increase of the MP2/TZVP entropies at temperatures larger than $T = 340$ K. In addition, it is apparent that the basis set effects found for the calculated entropies are not reflected in the accuracies of the corresponding isobars listed in Table 4.4. The largest basis set effect of approximately 10 J/(mol K) in the high temperature domain is clearly observed for the entropies obtained from the MP2 cluster calculations, whereas the BP functional shows the largest sensitivity with regard to this point in the case of the accuracies of the computed isobars.

However, the most significant deviations over the whole temperature interval are found for the entropies calculated on the basis of the MP2/TZVPP pair interaction energies, which amount to approximately 8 J/(mol K) at elevated temperatures. Compared to the even larger discrepancies with regard to the entropies of the fully cooperative cluster interaction energies found in the parameter-free setup as well as in the unmodified model (see Figs. 4.13 and 4.16), this is a considerable improvement and agrees to the increased accuracy observed for the MP2/TZVPP pair isobar obtained in the frame of the parameter adjustment procedure. These results thereby indicate that possible shortcomings in the supramolecular cluster interaction energies can be compensated to a certain degree via the intercluster mean field interaction not only in the case of the molar volume but also for the liquid phase entropy. However, it is again apparent that the entropies obtained from the fully cooperative cluster interaction energies are more accurate in almost all cases and at almost all temperatures. The liquid phase entropies calculated for the extended water cluster set on the basis of the optimized qce parameters from Table 4.4 are illustrated in Fig. 4.19 together with the experimental entropy of liquid water. It is clearly seen that the accuracy as well as the general progression of the calculated entropies is affected to virtually no extent by the addition of the spiro cluster structures in the case of the MP2/TZVP as well as the MP2/TZVPP pair cluster data, whereas larger deviations are obtained for all dft methods. These observations indicate that even though there is no direct correlation between the quantitative performance of the calculated molar volumes and the quality of the corresponding entropies, a rough estimate of the accuracy of the qce entropy can be obtained from the $\|\Delta \mathbf{V}\|$ values in this case, see Table 4.4. The additional consideration of the spiro structures results in a considerable loss of accuracy in the case of the B3LYP/TZVP isobar, and the entropies obtained from the B3LYP/TZVP cluster calculations also show the largest discrepancies with regard to the experimental reference in the frame of the extended water cluster set. The remaining dft approaches all predict very similar entropies over the whole temperature interval, which are more accurate by approximately 3 J/(mol K) at

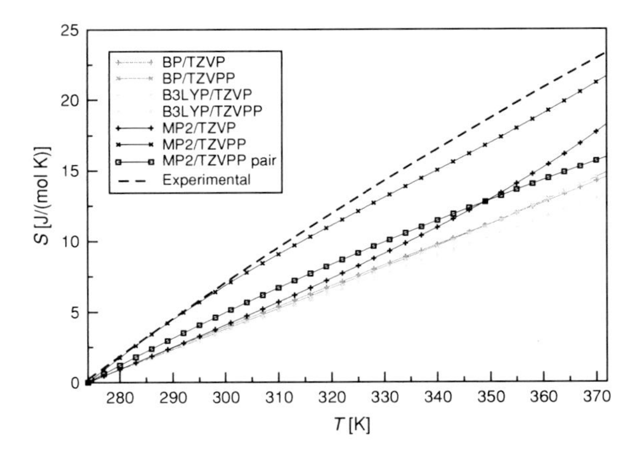

Fig. 4.19 Molar entropy S as a function of the temperature T as obtained from different quantum chemical methods employed for the calculation of the extended (7w8cube + spiro) cluster set compared to the experimental entropies [61]. The reference for all curves is set to $S(273.15\ \text{K}) = 0$ J/(mol K). The qce parameters are fitted to reproduce the experimental volume (see Table 4.4). All calculations refer to a pressure of $p = 1 \times 10^5$ Pa

elevated temperatures than the B3LYP/TZVP results, but exhibit a loss of accuracy of approximately 5 J/(mol K) as compared to the corresponding entropies obtained from the 7w8cube cluster set. This result also shows that the consideration of the additional cluster structures considerably reduces the basis set effect in the entropies of the dft cluster calculations, whereas a lesser influence of this kind can be found in the case of the MP2 computations. In contrast to the situation observed for the 7w8cube set, all MP2 results underestimate the experimental entropy in the case of the extended cluster set, which is the result of a shift in the MP2/TZVPP entropy curve of approximately 7 J/(mol K) towards lower values. In agreement with the increased performance of the MP2/TZVPP isobar in the frame of the extended set (see Table 4.4), this shift leads to an improved accuracy of the corresponding entropies over the whole liquid phase temperature interval with deviations smaller than 2 J/(mol K) at all considered temperatures. As in the case of the remaining methods, the deviations are considerably reduced at lower temperatures and become negligible for the values predicted from the MP2/TZVPP cluster data at temperatures smaller than $T = 300$ K. Thus, the extension of the 7w8cube cluster set in terms of the spiro structures generally results in a decrease of the liquid phase qce entropies, which again cannot directly be inferred from the isolated cluster entropies listed in Table 4.6, where these structures exhibit the largest individual values. In qualitative agreement with the changes found in the accuracies of the isobars for the transition from the 7w8cube to the extended cluster set, larger deviations of the calculated entropies are found for all dft methods, whereas the MP2/TZVP and MP2/TZVPP pair entropies are only affected in a negligible way and the accuracy of the MP2/TZVPP entropies is considerably increased. Thus, the previously discussed effect of a reduced accuracy due to the consideration of additional cluster structures is also apparent in the dft-based qce entropies, but the optimization of the cluster set with regard to clusters exhibiting small populations can improve this situation also in the case of the calculated entropies [2, 3]. However, in the case of the cluster calculations based on the most sophisticated of all considered quantum chemical methods

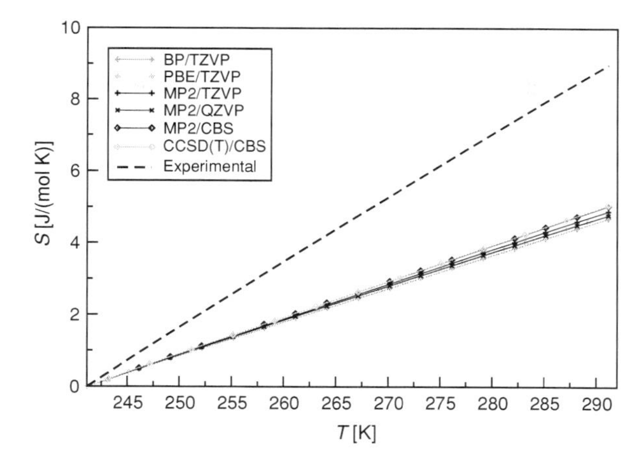

Fig. 4.20 Molar entropy S as a function of the temperature T as obtained from different quantum chemical methods employed for the calculation of the hydrogen fluoride cluster set (see Fig. 4.3) compared to a liquid phase entropy curve obtained from heat capacity measurements [55]. The reference for all curves is set to $S(241.15\ \text{K}) = 0\ \text{J/(mol K)}$. The qce parameters are fitted to reproduce the experimental volume (see Table 4.5). All calculations refer to a pressure of $p = 1 \times 10^5$ Pa

(MP2/TZVPP), the addition of the spiro clusters clearly leads to an improvement of the liquid phase entropies (see Fig. 4.19), which again indicates that the quality of the results obtained from the qce approach is very sensitive to these details of the employed cluster data. The entropies calculated for the upper half of the liquid hydrogen fluoride temperature interval on the basis of the optimized parameters (see Table 4.5) are illustrated in Fig. 4.20. It should be noted that the curve labeled as "Experimental" is only an estimate of the real entropy of liquid hydrogen fluoride (for which no experimental data could be found in the literature) based on a simple trapezoid rule integration scheme of experimentally determined heat capacities [66]. Therefore, the accuracy of this reference is difficult to estimate. With regard to the reference entropy obtained in this way, it is apparent that all employed methods predict entropies being smaller by approximately 3–4 J/(mol K), which is in agreement to the deviations found in the case of the extended water cluster set, see Fig. 4.19. It is also clearly seen that all quantum chemical methods applied for the generation of the cluster data yield very similar entropies in the examined temperature interval with virtually no differences between the various methods at temperatures lower than $T = 255$ K and differences smaller than 1 J/(mol K) in the high temperature domain. This finding is in contrast to the molar volumes obtained from the parameter adjustment procedure (see Fig. 4.12), for which considerable differences between the MP2/TZVP and the MP2/QZVP isobar and the curves calculated on the basis of the remaining methods are observed. As in the case of the entropies calculated for the extended water set (see Fig. 4.19), the results based on the dft cluster data assume the smallest entropy values at all temperatures and thereby underestimate the experimental reference to the largest extent. This trend is most pronounced in the entropies obtained from the BP/TZVP combination, whereas the PBE/TZVP results show a nearly quantitative agreement to the curve based on the MP2/QZVP cluster calculations. With regard to the results obtained from the parameter adjustment procedure of the qce isobars (see Table 4.5 and Fig. 4.12), it is apparent that in this case a correlation between the accuracies $\|\Delta \mathbf{V}\|$ and the precision of the calculated entropies is not reasonable, i.e., the isobar obtained from the PBE/TZVP cluster data is the most precise one of all employed methods, whereas the corresponding MP2/QZVP curve shows very large discrepancies to the experimental reference, see Fig. 4.12. It is also clearly visible that the basis set effects on the calculated entropies for the MP2 method are considerably reduced as compared to the ones found for the qce entropies of liquid water, see Figs. 4.18 and 4.19. Furthermore, these effects do not show a uniform trend with regard to the basis set size, i.e., the calculated entropies decrease for the transition from TZVP to QZVP and increase for the transition from QZVP to the complete basis set limit. The curves obtained from the MP2 and CCSD(T) methods at the complete basis set limit are again virtually identical, which is in agreement to the results found for the idealized parameter setups (see Figs. 4.14 and 4.17) as well as for the molar volumes (see Figs. 4.7 and 4.12). As in the case of the entropies calculated for the extended water cluster set (see Fig. 4.19), the methods exhibiting the most elaborate treatment of electron correlation for the hydrogen fluoride cluster calculations also yield the most accurate entropy curves with regard to the experimental reference. However, an assessment of the real precision of the calculated entropies is difficult considering possible inaccuracies in the experimental curve plotted in Fig. 4.20. A literature value of $S_{\mathrm{exp}}(292.69\ \mathrm{K}) = 74.5$ J/(mol K) for the absolute entropy of liquid hydrogen fluoride at the boiling point based on calorimetric measurements indicates that the corresponding entropies obtained from the adjusted parameter qce calculations are in reasonable agreement, with deviations again amounting up to 4 J/(mol K) for the most accurate electronic structure methods ($S_{\mathrm{CCSD(T)/cbs}}(291.15\ \mathrm{K}) = S_{\mathrm{MP2/cbs}}(291.15\ \mathrm{K}) = 70.9$ J/(mol K), $S_{\mathrm{PBE/TZVP}}(291.15\ \mathrm{K}) = 66.4$ J/(mol K), $S_{\mathrm{BP/TZVP}}(291.15\ \mathrm{K}) = 65.6$ J/(mol K)) [56].

Thus, the results presented in this section show that the parameters obtained from the adjustment procedure of the calculated isobars are also capable of predicting the entropy of liquid water and liquid hydrogen fluoride to within 5–10 J/(mol K) in most cases. For the results obtained from the extended water cluster set, a rough estimate of the quality of the calculated entropies can be gained from the $\|\Delta \mathbf{V}\|$ values of the parameter adjustment procedure, but an analogous correlation cannot be found in the case of the entropies calculated for the smaller 7w8cube set and the hydrogen fluoride cluster set. With the exception of the entropies obtained from the 7w8cube cluster set, it is again found that the most sophisticated electronic structure methods also yield the most accurate liquid phase entropies if the applied basis set is of sufficient size, but the MP2 hydrogen fluoride results demonstrate that there is no systematic improvement in the entropies with regard to the dimension of the applied basis set. In addition, the observed trends do not correlate with the calculated interaction energies (see Tables 4.1, 4.2, and 4.3), thereby indicating an increased importance of the calculated frequencies for the qce entropies. However, due to the iterative determination of the populations and the volume which both enter the entropy expression according to Eq. 2.68, a quantification of these effects cannot be accomplished in a straightforward way.

4.2.3 Summary of Liquid Phase Calculations

The results of the present chapter show that the modeling of thermodynamic liquid phase behavior in principle is possible on the basis of static first principles cluster calculations, thereby indicating that for the treatment of the condensed phase the qce approach is a useful extension of the conventional rrho model. The latter has been found to be problematic in the preceding chapter especially for entropy calculations if no modifications are taken into account. With regard to this point it is important to note that the qce approach (like the rrho model) relies on input data readily obtainable from static first principles calculations (with the exception of the two qce parameters), which permits the application of high quality post-Hartree–Fock methods and thereby the examination of substances exhibiting a complicated electronic structure over a rather wide temperature interval. This is not possible in most md simulation approaches, which would be the natural choice for the modeling of the condensed phase. However, there are also some problematic issues that have to be regarded for the general applicability of this method. In accordance with the final part of Chap. 3, this section will summarize the most important points of the results obtained in the frame of the qce calculations for the liquid phases of water and hydrogen fluoride.

- Static first principles calculations of small to medium sized hydrogen fluoride and water clusters employing dft methods (BP, B3LYP, PBE) as well as post-Hartree–Fock approaches (MP2, CCSD(T)) in combination with triple-ζ (TZVP, TZVPP) and quadruple-ζ (QZVP) basis sets as well as an extrapolation to the complete basis set limit (cbs) yield supramolecular interaction energies in the order of magnitude of several tens of kJ/mol for the smaller clusters and several hundreds of kJ/mol for the larger structures, see Tables 4.1, 4.2, and 4.3. The calculation of pairwise interaction energies (see Tables 4.1 and 4.2) and the analysis of the degree of charge transfer via the hydrogen bond in terms of the occupancy of the involved $\sigma^{\star}$ orbitals (see Fig. 4.4) indicate a high degree of cooperativity in the cyclic cluster structures.
- The combination of moderately sized cluster sets (see Figs. 4.1, 4.2, and 4.3) with a complete parameter-free qce setup ($a_{\mathrm{mf}} = 0$ (J $\times$ m^3), $b_{\mathrm{xv}} = 0$) shows the typical ideal gas behavior in the high temperature domain, but at certain temperatures (which are considerably dependent on the quantum chemical

methodology employed for the cluster calculations) a condensation to a high density liquid-like phase is observed, see Figs 4.5, 4.6, and 4.7. This condensation phenomenon only occurs if the supramolecular cluster interaction energies consider cooperative effects (see e.g. Fig. 4.6), and even though the obtained phase transition temperature in general differs from the experimental one to a large extent, the application of sophisticated quantum chemical methods for the cluster calculations can improve the situation (see Fig. 4.7). The consideration of an unscaled excluded volume estimate according to the atomic van der Waals radii ($b_{xv} = 1$) does not affect the results in any qualitative way. The high density phases obtained from this parameter setup exhibit molar volumes which are approximately two orders of magnitude larger than the corresponding experimental values.

- The situation completely changes if residual cluster–cluster interactions are considered in terms of the unscaled intercluster mean field interaction, see Figs. 4.8 and 4.9. The calculated molar volumes drop by approximately two orders of magnitude, but with regard to the experimental isobars the calculated values are too small by up to a factor of 2/3 in the case of hydrogen fluoride. In addition, convergency problems start to show up for certain cluster set configurations (the extended water cluster set) and certain temperatures (e.g. in the high temperature domain of the hydrogen fluoride BP/TZVP isobar), which have not been observed before in the case of the parameter-free setup to this extent. The consideration of pairwise cluster interaction energies as well as different combinations of dft-based supramolecular interaction energies, principal moments of inertia, and harmonic frequencies demonstrate that the isobars obtained within the frame of the mean field approach considerably depend on the employed interaction energies and on the harmonic frequencies, see Figs. 4.8 and 4.9.
- A simple least-squares procedure employed for the adjustment of the two qce parameters to experimental reference isobars (see Tables 4.4 and 4.5) significantly improves the accuracies of the calculated molar volumes and leads to average deviations $||\Delta \mathbf{V}||$ of below 3%. Larger deviations and convergency problems are found in the low temperature part for the isobars predicted by the 7w8cube cluster set and the hydrogen fluoride cluster set, see Figs. 4.10 and 4.12. The low temperature accuracy of the results based on the 7w8cube set can be improved by additionally including the spiro cluster structures in the qce calculations, thereby indicating that the fourfold coordination is an important motif at these temperatures, see Fig. 4.11. The improved performance of the isobar calculated from the pair interaction energies indicates that deficiencies in the cluster interaction energies can be compensated by the mean field approach for this particular case, but the larger discrepancies found for the MP2/TZVP and MP2/QZVP isobars of hydrogen fluoride demonstrate that such a correction is not generally possible, see Figs. 4.10 and 4.12. In the case of the extended water cluster set, the most accurate isobar is obtained from the most elaborate treatment of the electronic structure in the cluster calculations (MP2/TZVPP, see Table 4.4), whereas the dft approaches as well as the post-Hartree–Fock

methods at the basis set limit show a similar performance in the case of hydrogen fluoride, see Table 4.5.

- The progression of the entropies calculated in the frame of the parameter-free qce model is qualitatively similar to the one of the molar volumes in this parameter setup (see Figs. 4.13 and 4.14) but differs considerably from the isolated cluster entropies predicted by the conventional rrho model (see Table 4.6), which again indicates that a reduction of the volume available for translational motion as obtained from the cluster equilibrium according to Eq. 2.63 is mandatory to model liquid phase behavior. The agreement between the entropies calculated from the parameter-free model and the experimental entropies of liquid water and hydrogen fluoride is much higher as in the case of the corresponding molar volumes, and the consideration of the unscaled inter-cluster mean field interaction ($a_{mf} = 1$ (J $\times$ m^3)/mol) leads to entropies which are already in reasonable agreement with the experimental reference, see Figs. 4.16 and 4.17. Differences between the various quantum chemical methods applied for the cluster calculations in general do not exceed 10 J/(mol K) with the exception of the entropies predicted by the MP2/TZVPP pair interaction energies, which are larger by approximately 30 J/(mol K) over the whole examined temperature interval, see Fig. 4.16.
- The liquid phase entropies predicted by the adjusted parameter qce setup agree to the experimental values within 10 J/(mol K) in all cases and always underestimate the experimental reference with the exception of the MP2/TZVPP data in the case of the 7w8cube set. Large basis set effects are found for the entropies calculated on the basis of the 7w8cube cluster set, which are considerably reduced in the case of the extended water cluster set and are nearly absent in the case of the hydrogen fluoride set, see Figs. 4.18, 4.19, and 4.20. Most accurate results are obtained from the most sophisticated electronic structure methods employed for the cluster calculations in the case of the extended water cluster set (MP2/TZVPP) and the hydrogen fluoride cluster set (CCSD(T)/cbs), whereas the entropies predicted by the B3LYP hybrid functional cluster data show the highest agreement with the experimental reference in the case of the smaller 7w8cube cluster set. However, no clear trend in the liquid phase entropies with regard to the applied methodology or the basis set size can be observed, and due to the large number of quantities the qce entropy depends on according to Eq. 2.68 (cluster populations and the qce volume in addition to the cluster properties already required in the conventional rrho approach) a systematic classification of the different influences is not straightforward.

An important result of the present chapter is the observation that the partial consideration of intermolecular interactions by embedding the isolated molecule into a cluster surrounding *and* the subsequent application of a chemical equilibrium between the various clusters is sufficient for obtaining a qualitative condensed phase behavior. The van der Waals-like parameters appearing in the qce approach help to adjust the calculated thermodynamic properties in a semi-quantitative way, but the basic transition from the ideal gas-like behavior of the rrho approach to the condensed phase is realized by replacing the single molecule by a set of clusters, i.e., by allowing several particles to interact with each other. The consideration of cooperative effects in these interactions is found to be of considerable importance in the presented results as well as in previous studies on this subject [1, 2, 15, 27, 59].

Thus, the calculation of liquid phase properties in the frame of the qce model has been shown to be generally possible to a reasonable precision in the present chapter. In contrast to most approaches relying on md simulation methods, the variation of the temperature interval does not constitute a significant bottleneck in qce calculations once the cluster data is available. Accordingly, the possibility of treating realistic phase transition processes via the appropriate extension of the temperature interval and a readjustment of the parameters should be principally possible, and the capability of this approach will be studied in the following chapter.

References

1. Spickermann C, Lehmann SBC, Kirchner B (2008) J Chem Phys 128:244506
2. Lehmann SBC, Spickermann C, Kirchner B (2009) J Chem Theory Comput 5:1640–1649
3. Lehmann SBC, Spickermann C, Kirchner B (2009) J Chem Theory Comput 5:1650–1656
4. Barker JA, Watts RO (1969) Chem Phys Lett 3:144–145
5. Rahman A, Stillinger FH (1971) J Chem Phys 55:3336–3359
6. Marx D, Hutter J (2009) Ab initio molecular dynamics: basic theory and advanced methods. Cambridge University Press, Cambridge
7. Leach AR (2001) Molecular modelling: principles and applications. Pearson Education, Essex
8. Franks F (1973) Water A comprehensive treatise. Plenum, New York
9. Ludwig R (2001) Angew Chem Int Ed 40:1808–1827
10. Wernet P, Nordlund D, Bergmann U, Cavalleri M, Odelius M, Ogasawara H, Naslund LA, Hirsch TK, Ojamae L, Glatzel P, Pettersson LGM, Nielsen A (2004) Science 304:995–999
11. Head-Gordon T, Johnson ME (2006) Proc Nat Acad Sci USA 103:7973–7977
12. Chatterjee S, Debenedetti PG, Stillinger FH, Lynden-Bell RM (2008) J Chem Phys 128:124511
13. Röntgen WC (1892) Ann Phys 281:91–97
14. Lee H, Tuckermann ME (2007) J Chem Phys 126:164501
15. Weinhold F (1998) J Chem Phys 109:373–384
16. Ludwig R (2007) ChemPhysChem 8:938–943
17. Klein ML, McDonald IR (1979) J Chem Phys 71:298–308
18. Jedlovszky P, Vallauri R (1997) J Chem Phys 107:10166–10176
19. Röthlisberger U, Parrinello M (1997) J Chem Phys 106:4658–4664
20. Valle RGD, Gazzillo D (1999) Phys Rev B 59:13699–13706
21. Kreitmeir M, Bertagnolli H, Mortensen JJ, Parrinello M (2003) J Chem Phys 118:3639–3645
22. Izvekov S, Voth GA (2005) J Phys Chem B 109:6573–6586
23. Deraman M, Dore J, Powles J, Holloway JH, Chieux P (1985) Mol Phys 55:1351–1367
24. McLain SE, Benmore CJ, Siewenie JE, Urquidi J, Turner JFC (2004) Angew Chem Int Ed 43:1952–1955
25. Pfleiderer T, Waldner I, Bertagnolli H, Tölheide K, Fischer HE (2000) J Chem Phys 113:3690–3696

26. Ludwig R, Weinhold F, Farrar TC (1995) J Chem Phys 103:3636–3642
27. Kirchner B (2005) J Chem Phys 123:204116
28. Ludwig R, Weinhold F (1999) J Chem Phys 110:508–515
29. Lenz A, Ojamäe L (2009) J Chem Phys 131:134302
30. Ludwig R, Weinhold F (2002) Z Phys Chem 216:659–674
31. Ludwig R, Behler J, Klink B, Weinhold F (2002) Angew Chem Int Ed 41:3199–3202
32. Wendt MA, Weinhold F, Farrar TC (1998) J Chem Phys 109:5945–5947
33. Borowski P, Jaroniec J, Janowski T, Woliński K (2003) Mol Phys 101:1413–1421
34. Ludwig R (2005) ChemPhysChem 6:1369–1375
35. Ludwig R (2005) ChemPhysChem 6:1376–1380
36. Song H-J, Xiao H-M, Dong H-S, Huang Y-G (2006) J Mol Struct (Theochem) 767:67–73
37. PEACEMAKER V 1.4 Copyright B Kirchner, written by Kirchner B, Spickermann C, Lehmann SBC, Perlt E, Uhlig F, Langner J, Domaros Mv, Reuther P, 2004–2009, University of Bonn, Institute of Physical and Theoretical Chemistry, University of Leipzig, Wilhelm-Ostwald Institute of Physical and Theoretical Chemistry Bonn-Leipzig 2009, see also http://www.uni-leipzig.de/ ~ quant/index.html/
38. Rincòn L, Almeida R, García-Aldea D, Riega HD (2001) J Chem Phys 114:5552–5561
39. Frank HS, Wen W (1957) Discuss Faraday Soc 24:133–140
40. Koßmann S, Thar J, Hunt PA, Welton T, Kirchner B (2006) J Chem Phys 124:174506
41. Reed AE, Curtiss LA, Weinhold F (1988) Chem Rev 88:899–926
42. Weinhold F (2006) Adv Protein Chem 72:121–155
43. Tomasi J, Mennucci B, Cammi R (2005) Chem Rev 105:2999–3093
44. Spickermann C (2006) Calculation of thermodynamical quantities in non-covalently bonded systems. Thesis, Rheinische Friedrich-Wilhelms-Universität Bonn
45. Cournoyer ME, Jorgensen WL (1984) Mol Phys 51:119–132
46. Liem SY, Popelier PLA (2003) J Chem Phys 119:4560–4566
47. Quack M, Stohner J, Suhm MA (2001) J Mol Struct 599:381–425
48. Stoll H (1992) Chem Phys Lett 191:548–552
49. Halkier A, Helgaker T, Jørgenson P, Klopper W, Koch H, Olsen J, Wilson AK (1998) Chem Phys Lett 286:243–252
50. Friedrich J, Perlt E, Roatsch M, Spickermann C, Kirchner B (2010) J Chem Theory Comput, submitted
51. Pine AS, Howard BJ (1986) J Chem Phys 84:590–596
52. Jensen F (1999) Introduction to computational chemistry. Wiley-VCH, Chichester
53. Guedes RC, do Couto PC, Cabral BJC (2003) J Chem Phys 118:1272–1281
54. Holleman AF, Wiberg N, Wiberg E (2007) Lehrbuch der Anorganischen Chemie. de Gruyter, Berlin
55. Yaws CL (ed) (1999) Chemical properties handbook. McGraw–Hill, New York
56. Vanderzee CE, Rodenburg WW (1970) J Chem Thermodyn 2:461–478
57. Geiger A, Kowall T (1994) Hydrogen bonding and molecular mobility in aqueous systems. In: Bellissent-Funel MC, Dore JC (eds) Hydrogen bond networks. Kluwer Academic Publishers, Dordrecht
58. Sun Q (2010) J Chem Phys 132:054507
59. Kirchner B (2007) Phys Rep 440:1–111
60. Silla E, Tunon I, Pascual-Ahuir JL (1991) J Comput Chem 12:1077–1088
61. National Institute of Standards and Technology, NIST chemistry webbook. see http://webbook.nist.gov/
62. Lide DR (2000) Handbook of chemistry and physics. CRC Press, Boca Raton
63. Simons JH, Bouknight JW (1932) J Am Chem Soc 54:129–135
64. McQuarrie DA, Simon JD (1997) Physical chemistry. University Science Books, Sausalito
65. Ludwig R, Weinhold F, Farrar TC (1997) J Chem Phys 107:499–507
66. Bronstein IN, Semendjajew KA, Musiol G, Mühlig H (2006) Taschenbuch der Mathematik. Verlag Harri Deutsch, Frankfurt am Main

Chapter 5
Phase Transitions

The final chapter of this thesis will focus on the calculation of changes in thermodynamic properties associated with the phase transition from the liquid to the gas phase. As in the previous chapters the primal quantity under investigation will be the (vaporization) entropy, but other phase transition properties will be computed as well. According to the results obtained from the parameter-free qce calculations for water and hydrogen fluoride as presented in the previous chapter, it is apparent that the transition from a high density liquid-like phase to the gas phase can be qualitatively modelled by this approach. In this chapter the possibility of calculating more accurate phase transition properties via a readjustment of the optimized qce parameters will be examined in detail, and an assessment with regard to values determined experimentally will be carried out. The systems employed for these investigations will again be the water and hydrogen fluoride cluster sets introduced in the previous chapter.[1]

5.1 Phase Transition Properties from the Parameter-Free Model

The isobars calculated in the frame of the parameter-free qce model exhibit the qualitative features of a liquid–vapor phase transition both in the calculations carried out for water as well as for hydrogen fluoride (see Figs. 4.5, 4.6, and 4.7), and a similar behavior is found in the case of the entropies obtained from the parameter-free setup. Even though there are differences of about two orders of magnitude between the molar volume of the high density phase predicted by this idealized qce setup and the molar volume of real liquid water, the changes in thermodynamic quantities in going from this high density phase to the gas phase

[1] Parts of this chapter have already been published in Refs. [1–3].

C. Spickermann, *Entropies of Condensed Phases and Complex Systems*, Springer Theses, DOI: 10.1007/978-3-642-15736-3_5,

could nevertheless show smaller discrepancies. Due to the relatively large temperature interval in which these transitions occur (see, e.g., Fig. 4.5), it is difficult to determine a precise condensation temperature at which the difference in the quantity under examination between the gas phase and the liquid-like phase is to be taken. Furthermore, it is apparent e.g. from Fig. 4.6 that the termination of the condensation process is significantly affected by the quantum chemical method applied for the calculation of the cluster data, which can lead to differences in the onset of the phase transition of up to 50 K (see, e.g., the curves based on the BP/TZVP and the MP2/TZVP combination, respectively, in Fig. 4.6). In order to overcome these difficulties, a uniform temperature of $T = 176$ K for water and $T = 200$ K in the case of hydrogen fluoride is chosen for the computation of the differences in the considered thermodynamic quantities between the high density phase and the gas phase. It is apparent from Figs. 4.6 and 4.7 that at these temperatures all applied methods predict the high density phase to be the thermodynamic stable one and that the condensation process is completed in all cases. The transition properties obtained in this temperature domain will not reflect the real first order liquid–vapor phase transitions of water and hydrogen fluoride, but a comparison of the calculated order of magnitude to the experimental values is nevertheless instructive.

5.1.1 Entropy Changes

It is apparent e.g. from the plots in Figs. 4.13 and 5.1 that in the case of water the phase predicted to be stable by the qce iterations at $T = 176$ K is the low density liquid-like phase. In order to obtain an estimate of the qce gas phase entropy at this temperature, a linear regression of the entropy values calculated for the high temperature domain has been carried out on the basis of the MP2/TZVP gas phase entropies at $T = 250$ K and all larger temperatures.

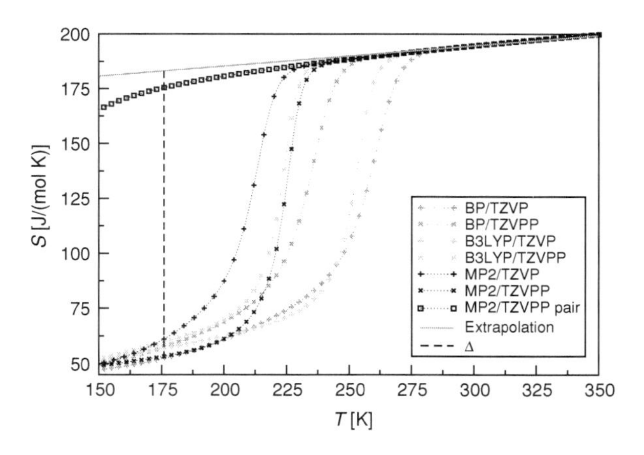

Fig. 5.1 Molar entropy S as a function of the temperature T as obtained from different quantum chemical methods employed for the calculation of the extended (7w8cube + spiro) cluster set. The label "Extrapolation" indicates the result obtained from the linear regression of the MP2/TZVP high temperature entropies. All calculations refer to a pressure of $p = 1 \times 10^5$ Pa

Table 5.1 Phase transition entropies at $T = 176$ K(H_2O ext.) and $T = 200$ K (HF) for the transition from the high density phase to the gas phase as obtained from the parameter-free qce setup (see Figs. 4.6 and 4.7) for the extended water cluster set (H_2O_{ext}) and the hydrogen fluoride cluster set (HF)

Set	dft Methods				
	BP		B3LYP		PBE
	TZVP	TZVPP	TZVP	TZVPP	TZVP
H_2O_{ext}	130.7	125.6	126.2	123.3	–
HF	108.2	–	–	–	105.8
	Post-Hartree–Fock methods				
	MP2				CCSD(T)
	TZVP	TZVPP	QZVP	cbs	cbs
H_2O_{ext}	122.0	129.8	–	–	–
HF	98.5	–	101.6	101.9	101.9

All values are given in [J/(mol K)]

The data based on the MP2/TZVP cluster calculations has been chosen for this approach, because this combination predicts the lowest temperature for the onset of the condensation process in the parameter-free setup (see Figs. 4.13 and 5.1) and thereby yields the largest data set for the linear regression. The result of this procedure is illustrated in Fig. 5.1 exemplarily for the extended water cluster set. The point at which the phase transition entropies are calculated is indicated via the label "Δ". Due to the poor convergence behavior of the 7w8cube cluster set in the frame of the parameter-free model in the low temperature domain (see Fig. 4.5), the calculation of phase transition entropies in this setup has been carried out only for the extended water cluster set and the hydrogen fluoride cluster set. The values obtained in this way are summarized in Table 5.1.

It is apparent that in the case of water all applied methods predict phase transition entropies between $\Delta S = 120\,\mathrm{J/(mol\ K)}$ and $\Delta S = 130\,\mathrm{J/(mol\ K)}$. The largest value is obtained on the basis of the BP/TZVP cluster calculations, which is no surprise considering the observation that this method predicts the lowest absolute entropy in the high density phase, see Fig. 5.1. With regard to this point it is also obvious why the smallest phase transition entropies are predicted by the MP2/TZVP and the B3LYP/TZVP cluster data, and that the situation would be different if another reference temperature (e.g. $T = 150$ K) would be chosen. There is no phase transition entropy for the MP2/TZVPP pair interaction energies listed in Table 5.1, and it is clear from Fig. 5.1 that at $T = 176$ K the stable phase in this case still is the gas phase and that the condensation process is shifted to even lower temperatures. The comparison of the values obtained from this parameter-free setup to the experimental vaporization entropy of real water at $T = 298.15\,\mathrm{K}$ ($\Delta_{vap}S(298.15\ \mathrm{K}) = 118.8$ J/(mol K)) indicates that the difference in entropy calculated between the two qce model phases is in the same order of magnitude like the one between the liquid and the vapor phase of real water. This is of course only a qualitative comparison due to the large difference in temperature of more than 100 K, but it shows that the entropic features of the real liquid–vapor phase transition are essentially captured in the qce model calculation and that a more accurate prediction could be possible via a readjustment of the qce parameters.

The phase transition entropies calculated for the hydrogen fluoride cluster set in the frame of the parameter-free setup are smaller than those obtained from the extended water cluster set and show a partitioning into a group of larger values predicted by the dft cluster data and a group of lower values found on the basis of the post-Hartree–Fock cluster calculations. Again this is the same ordering as already plotted in the low temperature part of Fig. 4.7, and according to these results identical phase transition entropies are predicted by both the MP2/cbs and the CCSD(T)/cbs cluster data. This almost quantitative equality between the results obtained from the MP2/cbs and the CCSD(T)/cbs calculations is also present in many results presented in the previous chapter and can be attributed to the fact that both approaches employ the same MP2/QZVP* cluster structures as well as the same harmonic frequencies and barely differ in the supramolecular interaction energies as can be seen from Table 4.3. In the case of the computed phase transition entropies in Table 5.1, the value found for the MP2/QZVP cluster data is also almost equal to the ones predicted by MP2/cbs and CCSD(T)/cbs as already indicated in the low temperature domain of Fig. 4.14. This again supports the observation that in the case of the hydrogen fluoride qce entropies the employed harmonic frequencies have a more significant impact than the supramolecular interaction energies (see e.g. Fig. 4.17), which is not found for the water cluster sets to this extent. In addition, it is also apparent from Fig. 4.14 that the effect of the reference temperature at which the phase transition entropy is evaluated is considerably smaller as in the case of water, i.e., even though the calculated values will quantitatively be different if this temperature is shifted by e.g. 50 K towards lower or higher values, the order of the different methods will be unaffected. If the entropy differences between the high density and low density hydrogen fluoride phases are compared to the experimental vaporization entropy of hydrogen fluoride at the boiling point ($\Delta_{\text{vap}}S$ (292.69 K) = 98.7 J/(mol K)), the same order of magnitude is again found for both quantities as already observed for the phase transition entropies of the extended water cluster set [4]. These results thereby again indicate that the modeling of the real liquid–vapor phase transition can possibly be achieved in the frame of the qce approach.

5.1.2 Other Quantities

As pointed out in Sect. 2.2.1, the purpose of the qce iterations lies in the determination of a self-consistent set of cluster populations and volumes from which the canonical partition function for the stable phase is calculated according to Eq. 2.46. Once this is accomplished, the calculation of thermodynamic quantities can be carried out in a straightforward analytical way as demonstrated in Sect. 2.2.2 using

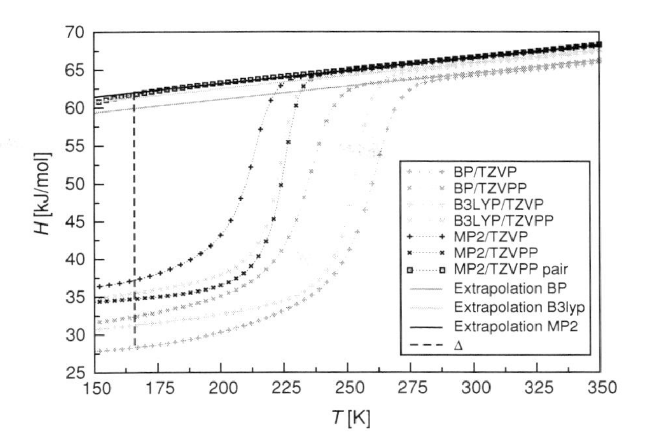

Fig. 5.2 Molar enthalpy H as a function of the temperature T as obtained from different quantum chemical methods employed for the calculation of the extended (7w8cube + spiro) cluster set. The label "Extrapolation" indicates the results obtained from the linear regression of the corresponding high temperature enthalpies. All calculations refer to a pressure of $p = 1 \times 10^5$ Pa

the example of the entropy. The major part of this thesis is focussed on the calculation of molar volumes and entropies, but other thermodynamic properties can be obtained from the N-particle partition function in an analogous manner. This course of action will be illustrated here exemplarily for the phase transition enthalpy and the change in the specific heat capacity computed for the transition from the high density phase to the gas phase as obtained in the frame of the parameter-free qce model. The enthalpies obtained from this parameter setup and the extended water cluster set are plotted exemplarily in Fig. 5.2.

In contrast to the entropies from Fig. 5.1 larger differences between the various methods employed for the cluster calculations are found, which amount to approximately 10 kJ/mol in the low temperature domain and still add up to several kJ/mol in the gas phase. The sequence in the enthalpies computed from the different quantum chemical methods in most cases follows the same trends as observed in the supramolecular cluster interaction energies of the larger cluster structures (see Tables 4.1 and 4.2), i.e., lowest enthalpies are obtained from the dft methods employing the smaller TZVP basis set and largest values are found for the MP2/TZVP combination. These clear trends are absent in the case of the corresponding entropies (see Fig. 4.13), and after the transition to the gas phase the observed correlation is no longer present. Again, the relatively large differences in the gas phase have not been found before neither in the case of the entropy nor the molar volumes, see e.g. Figs. 4.6 and 4.14, and it is apparent from Fig. 5.2 that largest deviations in the elevated temperature domain are found between the different methods rather than between the basis sets employed for a particular method. Accordingly, a linear regression for each of the two density functionals applied for the cluster calculations as well as for the MP2 data has been conducted separately employing the results of the larger TZVPP basis set in each case to account for the observed differences in the gas phase (see the labels "Extrapolation" in Fig. 5.2). Furthermore, the reference temperature for the calculation of the enthalpy difference between the two phases has been reduced to $T = 166$ K, thereby considering the large temperature interval for the transition process

Table 5.2 Phase transition enthalpies at $T = 166$ K(H_2O ext.) and $T = 200$ K (HF) for the transition from the high density phase to the gas phase as obtained from the parameter-free qce setup (see Figs. 4.6 and 4.7) for the extended water cluster set (H_2O_{ext}) and the hydrogen fluoride cluster set (HF)

Cluster set	dft Methods				
	BP		B3LYP		PBE
	TZVP	TZVPP	TZVP	TZVPP	TZVP
H_2O_{ext}	31.6	27.5	30.1	25.8	–
HF	32.4	–	–	–	35.3
	Post-Hartree–Fock methods				
	MP2				CCSD(T)
	TZVP	TZVPP	QZVP	cbs	cbs
H_2O_{ext}	24.8	27.2	–	–	–
HF	22.4	–	25.8	28.3	28.3

All values are given in [kJ/mol]

observed for the MP2/TZVP data, which indicates that the condensation to the high density phase is not completed at $T = 176$ K.

In the case of the hydrogen fluoride cluster set a similar behavior as for the extended water set is observed, i.e., there are differences of several kJ/mol in the gas phase enthalpies between the predictions obtained from the post-Hartree–Fock and the dft cluster data (not shown). However, the enthalpies calculated on the basis of the two considered density functionals (BP and PBE) as well as those predicted by the MP2 and the CCSD(T) computations are very similar in the high temperature domain, so that a single linear regression for the dft data and for the post-Hartree–Fock data is sufficient for a reasonable estimate of the phase transition enthalpies. The enthalpy changes obtained from these linear regressions are summarized in Table 5.2.

In the case of the extended water cluster set, analogous (but reversed) trends as already observed in the absolute enthalpies of the high density phase as well as in the supramolecular cluster interaction energies are found (see Tables 4.1 and 4.2), i.e., the phase transition enthalpies predicted by the BP/TZVP and the B3LYP/TZVP cluster data assume the largest values, whereas the smallest enthalpy change is obtained from the MP2/TZVP calculations. However, the observed differences do not exceed values of approximately 7 kJ/mol and thereby are considerably smaller as e.g. the variations found between the different methods employed in the calculation of the supramolecular interaction energies listed in Tables 4.1 and 4.2. As in the case of the entropies plotted in Fig. 5.1 the condensation process to the high density phase is not fullfilled for the enthalpy calculated from the MP2/TZVPP pair interaction energies even at $T = 150$ K, which is the reason why no value for this approach is listed in Table 5.2. The comparison of the calculated values to the vaporization enthalpy of real water at the boiling point ($\Delta_{vap}H(373.15\ \text{K}) = 39.5$ kJ/mol) indicates that the phase transition enthalpies obtained from the parameter-free qce setup are in the correct order of magnitude, even if a very large difference in the underlying reference temperature of more than 200 K is present [5]. This is in agreement with the results obtained from the parameter-free setup for the phase transition entropies, see Table 5.1.

The enthalpy changes calculated for the hydrogen fluoride cluster set can again be divided into a group of larger values predicted by the dft approaches and a group containing the values obtained from the post-Hartree–Fock methods which are smaller than those of the first group by at least 4 kJ/mol. The largest phase transition enthalpy is calculated for the PBE/TZVP cluster data in accordance with the observation that the supramolecular interaction energies obtained from this method predict the most stable cluster structures and thereby the strongest intermolecular interactions in the condensed phase, see Table 4.3. This behavior is also found in the case of the remaining methods, i.e., the correlation between strong intermolecular interactions in the high density phase and corresponding to that a large value of the phase transition enthalpy (and vice versa) is clearly reflected in the values listed in Table 5.2. Thus, the enthalpy changes calculated for the hydrogen fluoride cluster set are considerably affected by the degree of intermolecular interaction, whereas the corresponding phase transition entropies in Table 5.1 show a larger dependancy on the employed harmonic frequencies, see the discussion at the end of Sect. 5.1.1. This can also be seen for instance in the case of the phase transition properties obtained from the MP2/QZVP cluster calculations, which predict an almost identical entropy change as compared to the corresponding result of the cbs limit extrapolation, whereas a different enthalpy change is calculated for both approaches as can be seen from Table 5.2. In contrast to the behavior found for the phase transition entropies, the enthalpy changes computed for the extended water cluster set and the hydrogen fluoride cluster set assume similar values in the range between $\Delta H = 25$ kJ/mol and $\Delta H = 35$ kJ/mol in most cases. As pointed out previously, this order of magnitude replies to the experimental vaporization enthalpy of real water at $T = 298.15$ K. However, the vaporization enthalpy of liquid hydrogen fluoride at the boiling point is considerably smaller than the corresponding value of liquid water ($\Delta_{\mathrm{vap}}H(292.69\ \mathrm{K}) = 7.5$ kJ/mol) [4]. Thus, the phase transition enthalpies calculated for the hydrogen fluoride cluster set in the frame of the parameter-free qce model overestimate the vaporization enthalpy of real hydrogen fluoride and predict an enthalpic situation which is closer to the one found for water. The reason for the particular small vaporization enthalpy of real hydrogen fluoride has been attributed to its highly non-ideal vapor phase, in which a considerable fraction of molecules are found to be associated [4, 6]. According to these findings, a lesser number of intermolecular interactions has to be broken for the transition from the liquid phase to the gaseous state, which finally results in a smaller value of the vaporization enthalpy as compared to, e.g., water or other hydrogen-bonded liquids like methanol [5]. Considering this aspect, the reason for the relatively large discrepancies between the phase transition enthalpies computed from the parameter-free setup and the experimental vaporization enthalpy at the boiling point can be traced back to the fact that the low density gas phase predicted by the qce model is an ideal one, i.e., the cluster populations calculated for this

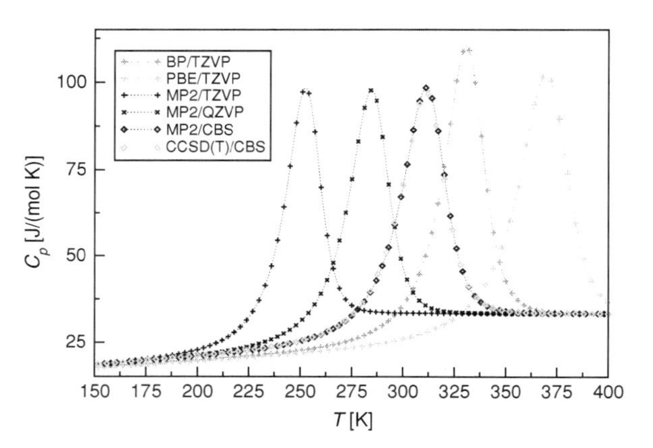

Fig. 5.3 Specific heat capacity C_p as a function of the temperature T as obtained from different quantum chemical methods employed for the calculation of the hydrogen fluoride cluster set (see Fig. 4.3). All calculations refer to a pressure of $p = 1 \times 10^5$ Pa

phase indicate a population of the hydrogen fluoride monomer of approximately 99% regardless of the method employed for the cluster calculations (not shown). In addition, there is also no contribution from the mean field intercluster interaction in the parameter-free setup, which means that the gas phase predicted by the qce calculations is essentially unassociated and more closely represents the circumstances found in the case of water. A possible correction of this situation in terms of the adjusted qce parameters will be examined in the following section.

As a second example of a thermodynamic quantity not considered so far the specific heat capacity at constant pressure is illustrated in Fig. 5.3 for the hydrogen fluoride cluster set calculated in the frame of the parameter-free setup.[2]

These plots again clearly show the pronounced two-phase behavior of the parameter-free qce setup in terms of distinctive turning points in the calculated curves as already observed in the case of the entropies (see Fig. 5.1), the enthalpies (see Fig. 5.2), or the molar volumes (see Sect. 4.1.2). The sequence of the turning points found for the heat capacities computed on the basis of the different cluster calculations agrees to the corresponding order of condensation temperatures in Fig. 4.7 as it is to be expected, and the most pronounced maximum is obtained from the BP/TZVP cluster data. In the low and high temperature domains all quantum chemical methods applied in the cluster calculations yield almost identical heat capacities, and the curves calculated from the MP2/cbs and the CCSD(T)/cbs cluster data are again virtually identical. However, it is also apparent that the parameter-free setup predicts a qualitatively wrong behavior of the C_p values in the high density phase for all employed methods, because the specific heat capacities of both hydrogen fluoride and water assume larger values in the liquid phase than in the gas phase [7]. According to the classical thermodynamic relation between the isobaric heat capacity and the temperature derivative of the enthalpy, this behavior also indicates inaccuracies in the slopes of the calculated enthalpy curves as e.g.

[2] Please note that the specific heat capacities calculated for the extended water cluster set and the parameter-free qce model qualitatively agree to the curves plotted in Fig. 5.3.

illustrated in Fig. 5.2 for the extended water cluster set. However, the heat capacities plotted in Fig. 5.3 are calculated directly from the qce partition function (see Eq. 2.46) and are not obtained e.g. as numerical derivatives of the calculated enthalpies. With regard to this point it should be noted that in contrast to the entropy and the enthalpy the heat capacity additionally depends on the second derivative of the N-particle partition function with respect to the temperature (as can e.g. be seen in Eq. 2.69 for the case of the isochoric heat capacity), and larger inaccuracies for thermodynamic quantities showing such a dependancy on Q have already been found in previous qce studies [8]. A detailed examination of this aspect indicates that these second derivatives with respect to the temperature are very sensitive to the frequencies employed in the vibrational partition function and that the consideration of anharmonic contributions can help to improve the situation in certain cases [8, 9]. Nevertheless, a qualitative correct behavior of the computed heat capacities can be observed for the temperature interval in which the phase transition occurs, and the comparison of the specific heat capacity calculated for the low density qce phase ($C_p \approx 33$ J/(mol K)) to the experimental value of gaseous hydrogen fluoride at $T = 298.00$ K ($C_p(298.00\text{ K}) = 29.14$ J/(mol K)) demonstrates that both numbers are in the same order of magnitude even though the intermolecular association found for the real gas phase is not predicted by the qce calculations.

Thus, the results of this section indicate that order-of-magnitude estimates for phase transition properties can already be obtained in the frame of the parameter-free model in many cases as long as the corresponding quantity does not depend on higher order derivatives of the N-particle partition function or special conditions as in the case of the gas phase association found for hydrogen fluoride apply. Even if the calculated curves do not reflect the corresponding experimental curves in a detailed way and the temperatures at which the phase transition properties are evaluated are not equal to the experimental phase transition points, a first estimate can be obtained in terms of the linear regression approach employed in this section. Following the course of the previous chapter, the determination of more precise phase transition properties in terms of a readjustment of the qce parameters will be examined in the next section.

5.2 Liquid–Vapor Phase Transition and Cooperativity

The final section of this thesis will be concerned with the calculation of phase transition properties from an optimized parameter setup in the frame of the qce model. The results presented in this chapter so far indicate that a transition process from the qce high density phase to the qce gas phase naturally occurs in the parameter-free setup, and the calculated phase transition properties qualitatively agree to the corresponding experimental values of water and hydrogen fluoride in most cases. In the preceding chapter it could be demonstrated that the parameter adjustment process as introduced in Sect. 2.2.1 leads to a higher agreement between calculated and experimental quantities in the case of the liquid phase calculations (see e.g. Table 4.4 and Fig. 4.19), and the question arises if such an improved performance can also be obtained for the liquid–vapor phase transition. The proposed adjustment procedure employs an experimental reference isobar for the fitting of the qce parameters, which means that the model cannot be predictive with regard to this particular quantity if this adjustment procedure is applied. However, the liquid phase entropies calculated on the basis of the optimized parameters (see Figs. 4.19 and 4.20) clearly show that the prediction of other quantities is possible to a reasonable precision via this approach. If the first order liquid–vapor phase transition as well as the experimental phase transition temperatures can be modelled via a readjustment of the qce parameters, a possible error cancellation in the calculation of the phase transition properties as compared to the absolute quantities presented in the previous chapter could even result in more accurate values. The results obtained from the liquid phase qce calculations furthermore demonstrate that in the case of the molar volume and the entropy the neglect of many-body effects in the computed supramolecular interaction energies (as expressed in the MP2/TZVPP pair interaction energies, see Tables 4.1 and 4.2) can be approximately corrected in terms of an increased mean field contribution without a too large loss of accuracy (see e.g. Table 4.4), i.e., a qualitative correct behavior is also obtained from the pairwise interaction energies. However, both qce parameters do not depend on the temperature and thereby identical values are applied over the whole examined temperature domain. Thus, the same a_{mf} parameter has to account for the intercluster interaction in the gas phase (expected to be considerably smaller than in the condensed phase) *and* possible shortcomings of the intermolecular interactions computed in terms of the cluster approach in the liquid phase. In the case of the pairwise interaction energies these shortcomings are especially pronounced, and it will be interesting to see if a compromise between the amplified magnitude of the mean field interaction in the high density phase (as expressed in the large magnitude of the optimized a_{mf} parameter in this phase, see Table 4.4) and the conditions in the gas phase can be obtained.

5.2.1 Isobars in the Liquid–Vapor Phase Transition Domain

In order to consider the liquid–vapor phase transition process for the parameter adjustment procedure, the temperature interval has been extended to $T = 400$ K for the water calculations and to $T = 305$ K in the case of the hydrogen fluoride calculations. The isobars obtained from the adjustment procedure for the 7w8cube water cluster set are plotted in Fig. 5.4, and the corresponding qce parameters are listed in the first block of Table 5.3.

The plots in Fig. 5.4 demonstrate that all quantum chemical methods employed for the cluster calculations are capable of reproducing the experimental boiling point of liquid water in an accurate way, and that the calculated phase transitions

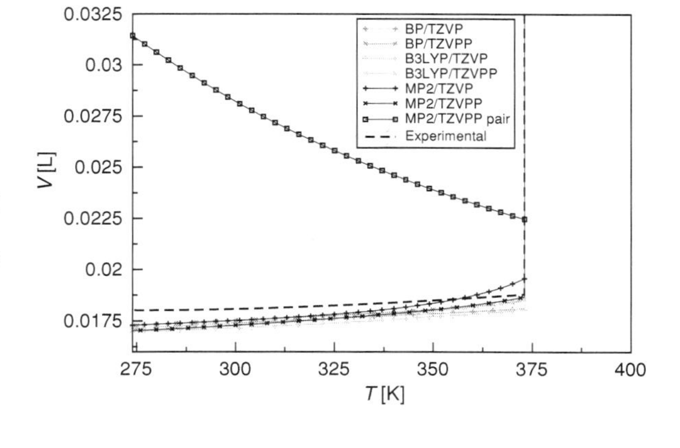

Fig. 5.4 Liquid phase section of the molar volume V as a function of the temperature T as obtained from different quantum chemical methods employed for the calculation of the 7w8cube cluster set compared to the experimental reference isobar [7]. The qce parameters are fitted to reproduce the experimental volume (see Table 5.3). All calculations refer to a pressure of $p = 1 \times 10^5$ Pa

Table 5.3 Optimized qce parameters a_{mf} and b_{xv} as well as the accuracies $\|\Delta\mathbf{V}\|$ for the liquid–vapor phase transition obtained from the adjustment procedure for both water cluster sets introduced in Sect. 4.1.1 $\|\Delta\mathbf{V}\|$ in [L] and a_{mf} in [(J $\times$ m^3)/mol]

Method	$\|\Delta\mathbf{V}\|$	a_{mf}	b_{xv}
7w8cube			
BP/TZVP	4.21	0.1000	1.030
BP/TZVPP	4.22	0.1000	1.030
B3LYP/TZVP	4.22	0.1000	1.030
B3LYP/TZVPP	4.24	0.1000	1.030
MP2/TZVP	4.20	0.1000	1.030
MP2/TZVPP	4.20	0.1000	1.030
MP2/TZVPP pair	4.20	0.1000	1.030
7w8cube + spiro			
BP/TZVP	4.21	0.1000	1.030
BP/TZVPP	4.22	0.1000	1.030
B3LYP/TZVP	4.22	0.1000	1.030
B3LYP/TZVPP	4.24	0.1000	1.030
MP2/TZVP	4.20	0.1000	1.030
MP2/TZVPP	4.20	0.1000	1.030
MP2/TZVPP pair	4.20	0.1000	1.030

now exhibit the rapid rise which is being characteristic for the first order phase transition in real water in contrast to the slow increase over an extended temperature interval observed in the case of the parameter-free model, see e.g. Fig. 4.6. Thus, a reasonable modeling of the liquid–vapor phase transition is in principle possible in the frame of the qce approach. All isobars predicted by the fully cooperative supramolecular cluster interaction energies agree qualitatively to the experimental curve and show only a small underestimation of the experimental liquid phase volume of approximately 0.6 mL at $T = 274$ K. These deviations are larger as compared to the accuracy of the isobars obtained from the parameter adjustment procedure in the liquid phase (see Fig. 4.10), but this behavior is to be expected due to the additional presence of the vapor phase in the adjustment process. In combination with the deviations found in the vapor phase section of the calculated isobars (see Fig. 5.5), the increased $\|\Delta\mathbf{V}\|$ values in Table 5.3 can be

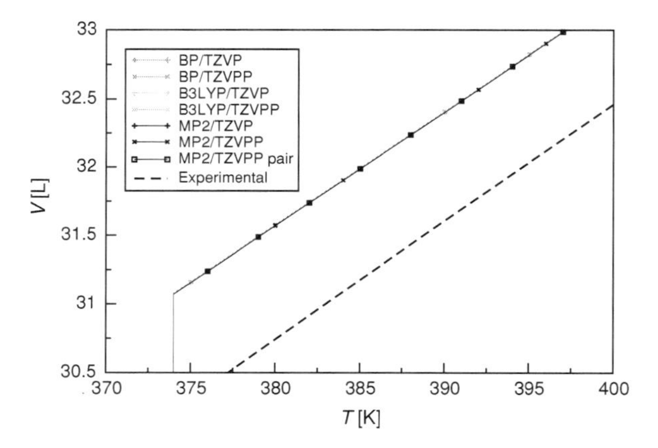

Fig. 5.5 Vapor phase section of the molar volume V as a function of the temperature T as obtained from different quantum chemical methods employed for the calculation of the 7w8cube cluster set compared to the experimental reference isobar [7]. The qce parameters are fitted to reproduce the experimental volume (see Table 5.3). All calculations refer to a pressure of $p = 1 \times 10^5$ Pa

rationalized as compared to the much smaller values listed in Table 4.4. It is also apparent from Fig. 5.5 that the major contribution to these increased mean square deviations arises due to the overestimation of the molar volume in the gas phase, whereas the deviations obtained from the liquid phase part of the calculated isobars are in the same order of magnitude as those found in the exclusive treatment of the liquid phase temperature interval. The smallest molar volumes in the liquid phase domain are calculated on the basis of the BP/TZVP and the B3LYP/TZVP cluster data at elevated temperatures, from which the most stable cluster structures and thereby strongest intermolecular interactions are obtained as well, see Table 4.1.

Intermediate (and very similar) molar volumes are predicted by all methods employing the TZVPP basis set at intermediate and elevated temperatures (with the exception of the MP2/TZVPP pair approach), whereas the small magnitude of the MP2/TZVP cluster interaction energies results in a larger overestimation of the experimental reference at elevated temperatures and a too early onset of the phase transition process. In the present cluster setup, the most accurate reproduction of the experimental molar volume at the boiling point is obtained from the MP2/TZVPP cluster calculations. However, the most striking deviations are clearly observed in the results calculated from the MP2/TZVPP pair interaction energies, which predict a qualitatively wrong progression of the molar volume over the whole liquid phase temperature interval. In addition to a considerable overestimation of the absolute magnitude by nearly a factor of two at $T = 274$ K, the curve based on the MP2/TZVPP pair interaction energies exhibits largest volumes at the low temperature end of the liquid phase interval and declines in a monotonic progression towards the boiling point (which is nevertheless correctly reproduced by these interaction energies). This trend is in qualitative contrast to the experimental data and thereby demonstrates that for a reasonable modeling of the liquid–vapor phase transition in the frame of the qce model the consideration of cooperative effects is necessary.

From the optimized qce parameters listed in Table 5.3 it is furthermore apparent that all applied quantum chemical methods yield almost identical values in the adjustment procedure of the liquid–vapor phase transition. With regard to this point it should be mentioned that the methodology applied for the parameter adjustment strongly disfavors isobars which do not match the phase transition temperature exactly, because a possible difference between calculated and experimental boiling point results in large contributions to the error vector $\Delta\mathbf{V}$ according to Eq. 2.53. Thus, the universal nature of the optimized parameters apparently is a prerequisite for the modeling of the experimental boiling point, at least in the case of the mean field interaction factor a_{mf}. This observation can be verified by test calculations, which show that a reduction of the mean field parameter by only 0.001 (J $\times$ m^3)/mol results in a drop of the calculated boiling point by 4 K in the case of the MP2/TZVPP 7w8cube cluster data. Comparing these numbers to the parameters optimized for the exclusive treatment of the liquid phase, larger variations are found in the latter data (see Table 4.4). In this case no phase transition point leading to a rapid change in the molar volume is present, thereby providing a greater degree of freedom for the parameters to account for details in the experimental reference curve.

With regard to this comparison it is also apparent that the parameters obtained for the phase transition and those of the liquid phase optimization employing the extended water cluster set are similar in most cases, and that the excluded volume estimate obtained from the atomic van der Waals spheres is a reasonable approximation to the excluded cluster volume in this case (as can be seen from b_{xv} values close to unity). The liquid phase section of the isobars calculated for the extended water cluster set on the basis of the optimized parameters in the second block of Table 5.3 are illustrated in Fig. 5.6. These plots indicate that the overall situation in the liquid domain of the extended temperature setup is not seriously affected by the addition of the spiro clusters. In general, all isobars are shifted towards lower volumes at liquid phase temperatures, and the B3LYP/TZVP cluster

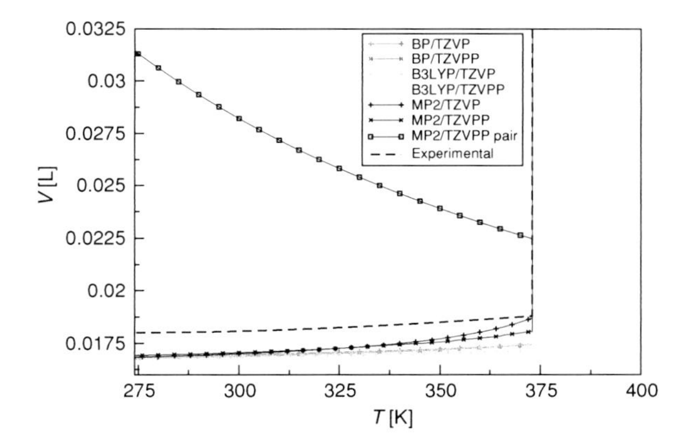

Fig. 5.6 Liquid phase section of the molar volume V as a function of the temperature T as obtained from different quantum chemical methods employed for the calculation of the extended (7w8cube + spiro) cluster set compared to the experimental reference isobar [7]. The qce parameters are fitted to reproduce the experimental volume (see Table 5.3). All calculations refer to a pressure of $p = 1 \times 10^5$ Pa

data predicts smallest volumes in the condensed phase as in the case of the 7w8cube cluster set. Due to the shift towards lower volumes the isobar calculated on the basis of the MP2/TZVP cluster data now yields the most accurate reproduction of the experimental molar volume at the boiling point. In the low temperature domain the variations in the calculated molar volumes between the different methods employed for the cluster calculations are less pronounced as compared to the results obtained from the smaller 7w8cube cluster set. In contrast to the situation observed in Fig. 5.4, there is virtually no basis set effect visible in the isobars obtained from the MP2 computations at low and intermediate temperatures, and in the temperature domain close to the boiling point both MP2 curves are also closer to each other than to the isobars based on the dft cluster calculations. This decreased difference between results obtained from the same method but a different basis set is also present in the case of the isobars predicted by the BP cluster calculations, and the same trend is observed for the curves calculated from the B3LYP cluster data though not as pronounced as in the previous cases. The only curve virtually unaffected by the extension of the employed cluster set is the isobar obtained from the MP2/TZVPP pair interaction energies, which exhibits the largest deviations with regard to the experimental reference and again shows a qualitative wrong behavior in the liquid phase temperature range. This result again indicates that in contrast to the situation found for the exclusive calculations of the liquid phase domain the neglect of cooperativity in the modeling of the liquid–vapor phase transition can neither be compensated via the extension of the cluster set in terms of coordination patterns possibly of special relevance at certain temperatures nor via an enlarged mean field interaction contribution. In addition, the discrepancies found between the isobar based on the MP2/TZVPP pair interaction energies and the experimental reference are considerably larger than those found for the different fully cooperative energies, which shows that in the case of water the consideration of cooperativity is more important for the liquid–vapor phase transition than e.g. a sophisticated treatment of electron correlation. These observations have already appeared in previous studies for the exclusive treatment of the liquid phase, but the present results demonstrate that cooperative effects are even more important for the qualitative correct treatment of the liquid–vapor phase transition in the frame of the qce model [10, 11].

The results presented so far show that a readjustment of the qce parameters with regard to the experimental boiling point yields qualitatively correct molar volumes if cooperative effects in the cluster interaction energies are considered as well as a quantitative reproduction of the experimental phase transition temperature. The liquid phase section of the isobars obtained from the optimized parameters are less accurate than those calculated from the parameters adjusted exclusively to the liquid phase due to the additional constraint imposed by the presence of the boiling point and the rapid change of the molar volume. As pointed out before, the methodology applied for the parameter adjustment favors those isobars which reproduce the phase transition temperature as close as possible. For the treatment of the liquid–vapor phase transition of hydrogen fluoride in the frame of the qce model an approach inverse to the procedure applied for the water cluster sets will be employed, i.e., the additional constraint of the phase transition point will be

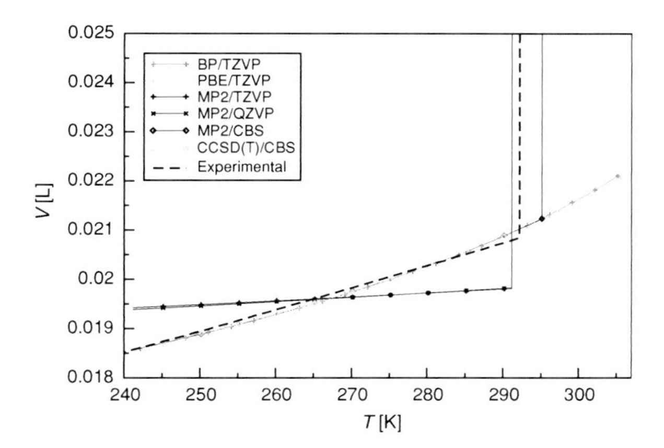

Fig. 5.7 Liquid phase section of the molar volume V as a function of the temperature T as obtained from different quantum chemical methods employed for the calculation of the hydrogen fluoride cluster set (see Fig. 4.3) compared to the experimental reference isobar [12]. The qce parameters are taken from Table 4.5. All calculations refer to a pressure of $p = 1 \times 10^5$ Pa

neglected and the performance of the liquid phase parameters as obtained in the previous chapter in predicting the phase transition temperature will be examined by simply extending the temperature range of the qce calculations to $T = 305$ K. Figure 5.7 illustrates the isobars calculated in the frame of this extended temperature interval employing the liquid phase parameters from Table 4.5 as well as the experimental isobar in the corresponding temperature range [12].

From these curves it is apparent that the experimental boiling point of hydrogen fluoride is predicted by all isobars which are based on the post-Hartree–Fock cluster calculations to within 4 K, whereas no liquid–vapor phase transition can be found in the case of the isobars obtained from the dft cluster data up to a temperature of $T = 305$ K. These findings indicate that the modeling of the liquid–vapor phase transition can be successful in the frame of the qce model on the basis of parameters which are adjusted according to a liquid phase reference only and that a readjustment of the qce parameters is not essential if minor deviations in the calculated phase transition temperature can be accepted.[3] With regard to this point it is furthermore apparent that the experimental boiling point is underestimated by the isobars based on the finite basis set MP2 cluster calculations, whereas it is overestimated by the MP2/cbs and the CCSD(T)/cbs cluster data. Considering the fact that the cluster properties obtained from the dft methods do not predict a phase transition up to $T = 305$ K, a correlation between the predicted boiling point and the strength of the intermolecular interaction as expressed in the cluster interaction energies (see Table 4.3) can be observed as in the case of the corresponding parameter-free qce isobars, see Fig. 4.7. However, the effect of these energies is much less pronounced as can be seen in the identical boiling temperatures predicted by the MP2/TZVP and the MP2/QZVP isobars and only leads to variations in the phase transition point in the order of magnitude of some Kelvin. Considering the combined accuracy of

[3] It should be noted that the analogous procedure does not result in a liquid–vapor phase transition for both the 7w8cube and the extended water cluster set in the extended temperature interval of up to $T = 400$ K.

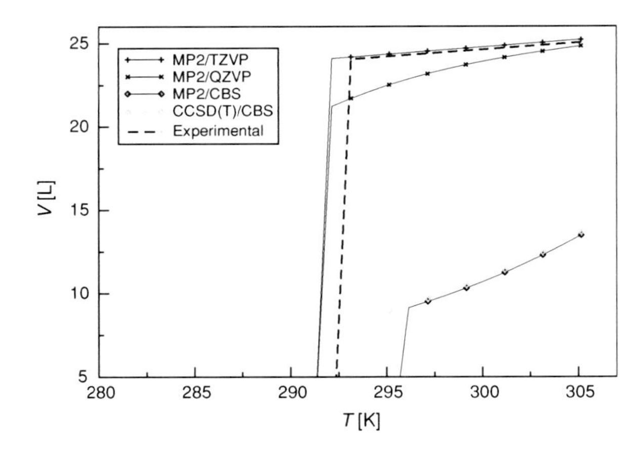

Fig. 5.8 Vapor phase section of the molar volume V as a function of the temperature T as obtained from different quantum chemical methods employed for the calculation of the hydrogen fluoride cluster set (see Fig. 4.3) compared to the experimental reference isobar [12]. The qce parameters are taken from Table 4.5. All calculations refer to a pressure of $p = 1 \times 10^5$ Pa

reproducing the liquid phase section *and* predicting the boiling temperature, it is found that the isobar obtained from the CCSD(T)/cbs cluster calculations shows the highest precision. Thus, the most elaborate treatment of the electronic structure again yields the most accurate result in the frame of the qce model as already observed previously (see e.g. Figs. 4.19 and 4.20), whereas the largest deviations are found for the curves based on the dft cluster data. However, this observation is only valid in the high density part of the calculated isobars. If the total change in molar volume for the transition to the gas phase is considered, the isobars predicted by the complete basis set limit extrapolations show considerable discrepancies to the experimental gas phase volume of real hydrogen fluoride. This behavior is illustrated in Fig. 5.8. From these plots it is seen that both isobars based on the complete basis set limit calculations yield a qualitatively wrong progression in the gas phase and underestimate the molar volume of real gaseous hydrogen fluoride by approximately 15 L, whereas the curve obtained from the MP2/QZVP cluster data also exhibits considerable discrepancies, but not to the extent observed for the MP2/cbs and CCSD(T)/cbs results. In contrast, the isobar computed from the MP2/TZVP cluster data and the liquid phase qce parameters predicts both the phase transition process as well as the molar volume of the gas phase in a qualitatively correct way. With regard to these findings it should be noted that the supramolecular cluster interaction energies obtained from the MP2/TZVP and the MP2/QZVP combinations yield the least stable cluster structures (see Table 4.3) and predict lowest condensation temperatures in the frame of the parameter-free model (see Fig. 4.7). Thus, it is to be expected that the relatively strong intermolecular interactions (as expressed in the large magnitudes of the interaction energies computed from the cbs limit extrapolations as well as the dft methods) in combination with a qce parameter setup optimized for the reproduction of the liquid phase molar volume cannot yield an adequate description of the gaseous phase at the same time. However, from Fig. 4.12 it is also clearly apparent that these strong intermolecular interactions are necessary to accurately reproduce the molar volume in the liquid phase, and that less stable clusters as computed on the basis of the MP2/TZVP and MP2/QZVP combinations are unable to reproduce the experimental curve accurately via an increased mean-field intercluster interaction as for instance found in the case of the MP2/TZVPP pair interaction energies in the qce water calculations (see e.g. Fig. 4.10). These observations thereby indicate that in the case of the hydrogen fluoride system a simultaneous high accuracy treatment of the liquid phase, the liquid–vapor phase transition, and the vapor phase does not seem possible in the frame of the qce model and that at most two of these three temperature regions can be precisely modelled at a time within the present approach. It should be noted that this situation cannot be changed by a readjustment of the qce parameters to the experimental molar volume in the phase transition domain as it has been carried out in the case of the water calculations (see Table 5.3).

This considerably different behavior of the hydrogen fluoride cluster set as compared to the calculated phase transition properties of the water cluster sets could be based on various reasons. First of all the consideration of only a single parameter setup for both the liquid as well as the vapor phase can be expected to constitute a severe restriction of the degree of intermolecular interaction. Considering the fact that both the supramolecular cluster interaction energies and the mean field parameter do not depend on the temperature, the only variation of the intermolecular interaction in the liquid phase and the gas phase can be accomplished in terms of the calculated volumes, see Eq. 2.51. However, the results calculated on the basis of the water clusters demonstrate that a qualitatively correct behavior of both phases can be obtained from this single parameter approach. Accordingly, a further examination of the calculated results indicates that the reason for the inability of the methods predicting the most stable cluster structures to additionally account for a qualitatively correct liquid–vapor phase transition can be found in the behavior of the isobars calculated in the frame of the parameter-free qce setup. The inspection of the curves obtained from the parameter-free setup (see Fig. 4.7) clearly show that the MP2/TZVP cluster data is the only one which predicts the gas phase as the stable phase at the experimental boiling temperature of $T = 292.69$ K and does not show an onset of the condensation process in this temperature region, whereas the remaining approaches either predict the high density phase as being most stable (BP/TZVP, PBE/TZVP) or exhibit molar volumes intermediate between the gas phase and the high density phase of the parameter-free model (MP2/QZVP, MP2/cbs, CCSD(T)/cbs) at this temperature. This means that with the exception of the MP2/TZVP cluster data all remaining approaches do not yield a characteristic gas phase behavior at the experimental boiling temperature even if the qce parameters are set to zero and no additional intermolecular interaction contribution is obtained from the mean field term, and it is obvious that such a behavior can even less be obtained from a parameter setup different from zero as applied for the calculation of the isobars in Figs. 5.7 and 5.8 With regard to these observations it is apparent that the only transition to a low density gas-like phase possible within the qce approach at a given temperature is the transition to the phase calculated from the parameter-free model at that temperature, and if the parameter-free model does not predict the gas phase to be the most stable one (or exhibits an intermediate behavior as in the case of the isobars obtained from the MP2/cbs and the CCSD(T)/cbs cluster data), no reasonable transition process will be observed regardless of the applied parameter setup. This is especially obvious in the case of the hydrogen fluoride isobars calculated on the basis of the dft cluster calculations in Fig. 5.7, which do not exhibit a phase transition point in the examined temperature interval at all. Compared to the corresponding curves obtained from the parameter-free model (see Fig. 4.7), it is apparent that the condensation point of the dft curves in this setup lies at approximately $T = 310$ K (BP) and $T = 350$ K (PBE), thereby lying outside of the examined temperature interval. The results of a test calculation for the BP/TZVP cluster data and an increased temperature interval of up to $T = 350$ K (not shown) employing the same qce parameters as the BP/TZVP isobar in Fig. 5.7 confirm this reasoning by predicting a phase transition point at approximately $T = 310$ K. Furthermore, this observation helps to explain why the qce calculations employing the water cluster sets successfully reproduce the qualitative features of the liquid–vapor phase transition of real water. The water isobars calculated in the frame of the parameter-free model (see Figs. 4.5 and 4.6) demonstrate that at the experimental boiling temperature of real water all methods employed for the cluster calculations predict the gas phase to be thermodynamically stable. Consequently, the isobars obtained from the parameter adjustment procedure can undergo a transition to this low density gas phase and thereby yield a behavior which is consistent with the experimental data.

The results of the present section thus show that the qce approach is capable of reproducing the characteristic features of a first order liquid–vapor phase transition and that a readjustment of the qce parameters is not essential for these calculations. The necessary precondition for a successful phase transition modeling is rather given by the behavior obtained from the parameter-free qce model at the experimental boiling point. The data obtained from the hydrogen fluoride calculations clearly show that the phase transition temperature can be predicted by employing parameters adjusted to liquid phase properties, but that a realistic modeling of the phase transition process is only possible if a gas phase behavior is predicted by the parameter-free model at the corresponding transition temperature. This observation is an important feature of the liquid–vapor phase transition modeling in the frame of the qce approach and demonstrates that the capability of the model to treat these transitions can be directly predicted by the behavior of the parameter-free results and that no time-consuming parameter adjustment is necessary in order to obtain this information. If the parameter-free results indicate that a vapor-like phase is stable at the transition temperature, qualitative correct results for the molar volume in the liquid, in the gas phase, and for the phase transition temperature can be expected as can be seen in the isobars obtained from the water cluster sets and the hydrogen fluoride isobar obtained from the MP2/TZVP cluster data, see Figs. 5.6 and 5.7. In the preceding chapter, the prediction of liquid phase entropies has been shown to be possible on the basis of the optimized qce parameters to a reasonable precision. The results of the following section will illustrate if this approach can also be employed for the prediction of vaporization entropies in the frame of the qce model.

5.2.2 First Principles Vaporization Entropies and Trouton's Rule

The entropy changes computed for the transition from the high density phase to the gas phase on the basis of the previously introduced cluster sets and the parameter-free qce model indicate that these values are in the right order of magnitude compared to the corresponding vaporization entropies of water and hydrogen fluoride, see Table 5.1. In the case of water, the qce parameters optimized to the molar volumes of the liquid–vapor phase transition and the boiling point (see Table 5.3) are employed in the present section for the assessment of the predictive capabilities of these parameters regarding the absolute entropies in the liquid and the gas phase as well as the vaporization entropy. The corresponding entropies obtained for the 7w8cube water cluster set in the extended temperature interval up to $T = 400$ K are plotted in Fig. 5.9. From this figure it is apparent that almost all methods employed for the generation of the cluster data predict the progression of the experimental entropy from the liquid to the vapor phase in a qualitatively correct way. The most accurate liquid phase entropies are calculated on the basis of the dft cluster data employing the smaller TZVP basis set, whereas the MP2/TZVP combination (along with the B3LYP/TZVP result) yields the most precise entropy value at the boiling point. Considering the whole liquid phase temperature interval, the largest basis set effects on the calculated entropies are again found in the MP2 cluster data, which is also observed in the entropies of the exclusive treatment of the liquid phase, see Figs. 4.18 and 4.19. Compared to these liquid phase calculations, it is again seen that the entropies computed from the MP2/TZVPP cluster data constantly overestimate the experimental values, whereas

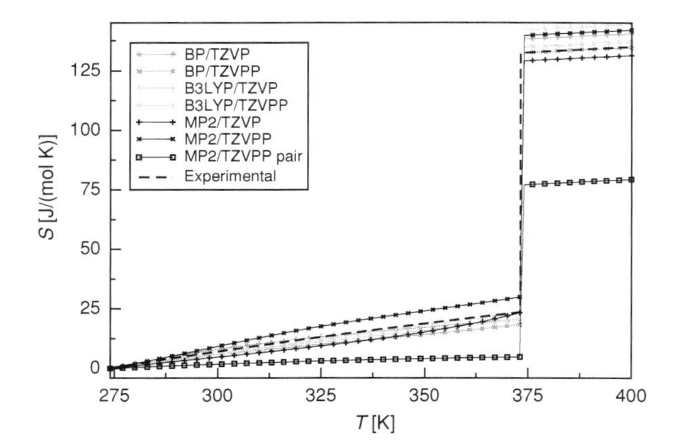

Fig. 5.9 Molar entropy S as a function of the temperature T as obtained from different quantum chemical methods employed for the calculation of the 7w8cube cluster set compared to the experimental entropies [7]. The reference for all curves is set to $S(273.15\text{ K}) = 0$ J/(mol K). The qce parameters are fitted to reproduce the experimental volume (see Table 5.3). All calculations refer to a pressure of $p = 1 \times 10^5$ Pa

the opposite behavior is found for the MP2/TZVP data. However, the most striking discrepancies are again observed for the entropies calculated on the basis of the MP2/TZVPP pair interaction energies, which predict an entropy contribution not larger than 5 J/(mol K) over the whole liquid phase temperature interval. With regard to the results obtained for the molar volumes in the liquid–vapor phase transition domain on the basis of these interaction energies (see Fig. 5.4), the situation has been improved insofar as the experimental entropies are predicted in a qualitatively correct way, but the large numerical deviations found for the MP2/TZVPP pair entropies indicate that the essential entropic conditions of the liquid water phase are not captured by this approach. This observation is also valid for the phase transition and the entropy calculated in the vapor phase.

Even though the phase transition point is accurately reproduced by the MP2/TZVPP pair interaction energies (as in the case of the molar volumes, see e.g. Fig. 5.4), the entropy predicted for the vapor phase again considerably underestimates the experimental values by more than 50 J/(mol K). In contrast, all curves computed from the fully cooperative interaction energies are in qualitative agreement to the experimental vapor phase entropies of water, with the BP/TZVPP combination showing the largest accuracy in this temperature domain. The values computed on the basis of the MP2/TZVP cluster data again underestimate the experimental curve, but this behavior is far from the considerable underestimation found in the case of the MP2/TZVPP pair interaction energies and amounts to a constant value of approximately 5 J/(mol K) only. All remaining methods predict gas phase entropies which are too large with regard to the experimental values by approximately 5–10 J/(mol K). With regard to this observation the largest overestimation is found for the B3LYP/TZVP cluster data, whereas the corresponding B3LYP/TZVPP entropies are considerably closer to the experimental curve. A basis set effect of similar extent is also seen in the entropies obtained from the MP2 and the BP cluster data, with the difference that the smaller basis set yields larger entropies in the case of the BP functional and smaller entropies in the case of the MP2 method. Comparing the accuracy of the calculated entropies in the different temperature domains, smaller deviations can be observed in the liquid phase interval for most methods, with the exception of the BP/TZVPP data yielding highly accurate values in the gas phase domain and the MP2/TZVPP entropies showing an almost constant overestimation in both phases. In agreement with the molar volumes calculated for the phase transition region (see Figs. 5.4 and 5.6) as well as to the results of the previous chapter, a prediction of entropies in the extended temperature interval is thus possible to a reasonable precision in both phases on the basis of qce parameters adjusted to the molar volume in the phase transition region. However, it is also apparent that the consideration of cooperative effects is a prerequisite in these calculations, and that in contrast to the situation observed for the exclusive treatment of the liquid phase the neglect of cooperativity cannot be compensated via the mean field interaction term.

The entropies obtained from the extended water cluster set in the liquid–vapor phase transition region are illustrated in Fig. 5.10. In general, the trends in the entropy observed for the extension of the water cluster set in terms of the spiro

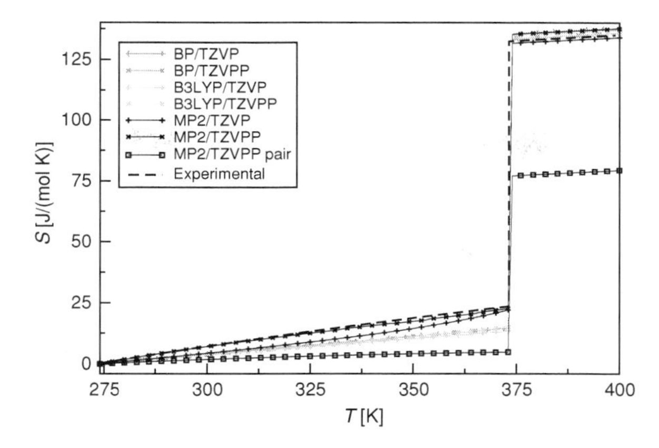

Fig. 5.10 Molar entropy S as a function of the temperature T as obtained from different quantum chemical methods employed for the calculation of the extended (7w8cube + spiro) cluster set compared to the experimental entropies [7]. The reference for all curves is set to $S(273.15\text{ K}) = 0$ J/(mol K). The qce parameters are fitted to reproduce the experimental volume (see Table 5.3). All calculations refer to a pressure of $p = 1 \times 10^5$ Pa

structures are comparable to the ones found in the exclusive treatment of the liquid phase, see Figs. 4.18 and 4.19. The various dft approaches employed for the cluster calculations again predict very similar entropies in the liquid phase temperature domain, and there are almost no basis set effects visible in these dft results. As in the case of the liquid phase computations (see Fig. 4.19), these methods also show a considerable underestimation of the experimental values in the liquid phase. In contrast, the entropies calculated on the basis of the MP2 cluster data are more accurate. This is especially apparent for the MP2/TZVPP data, which predicts the experimental entropies in the low temperature domain almost exactly and shows only negligible deviations at elevated temperatures. A similar behavior is found for the entropies from the MP2/TZVP cluster data near the phase transition temperature, but at lower temperatures there are larger deviations apparent as the MP2/TZVP curve approaches those obtained from the dft methods. At the boiling point, the most accurate entropies are predicted by the MP2 cluster data. The gas phase entropies calculated for the extended water cluster set show a considerable lesser degree of scattering as compared to the corresponding values obtained from the 7w8cube cluster set, and the MP2/TZVP cluster data again yields the only curve which underestimates the experimental vapor entropies. However, this underestimation (as well as the overestimation found for the remaining methods) is not as pronounced as in the case of the gas phase entropies of the smaller 7w8cube cluster set.

The most accurate predictions in this temperature domain are observed for the values based on the dft cluster data employing the TZVPP basis set, and the differences between the results of the two considered density functionals for a given basis set are smaller than those between the values obtained from the same density functional with varying basis sets. More important, it is obvious from Fig. 5.10 that the large discrepancies in the entropies predicted from the MP2/TZVPP pair interaction energies cannot be eliminated by additionally considering the spiro cluster structures. The entropy curve obtained from these energies is the only one which is virtually unaffected by the extension of the cluster set and thereby again demonstrates that the neglect of cooperative effects can neither account for the entropic situation in the liquid phase nor yield a reasonable description of the liquid–vapor phase transition process in entropic terms. Thus, the deficiencies introduced by the pairwise interaction energies affect the qce iterations in a fundamental way, and in contrast to the calculations exclusively covering the liquid phase interval this behavior cannot be corrected in terms of variations in the qce parameters or the cluster set according to the results presented in this chapter so far.

The entropies calculated for the hydrogen fluoride cluster set in the liquid–vapor phase transition region are illustrated in Fig. 5.11. The curve labeled as "Experimental" is again obtained from a numerical integration of experimentally determined heat capacities in the liquid phase temperature interval and the vaporization entropy at the boiling point ($\Delta_{\text{vap}}S(292.69\text{ K}) = 98.7$ J/(mol K)) added afterwards [4, 5]. Since the curves in Fig. 5.11 are not based on a renewed parameter adjustment procedure as carried out for the liquid–vapor phase transition calculations of water but instead are obtained from a simple extension of the temperature interval, the entropies predicted for the liquid phase temperature domain are identical to the ones in Fig. 4.20.

Accordingly, the liquid phase entropy is again underestimated by all methods employed for the cluster calculations, and there is almost no variation in the values

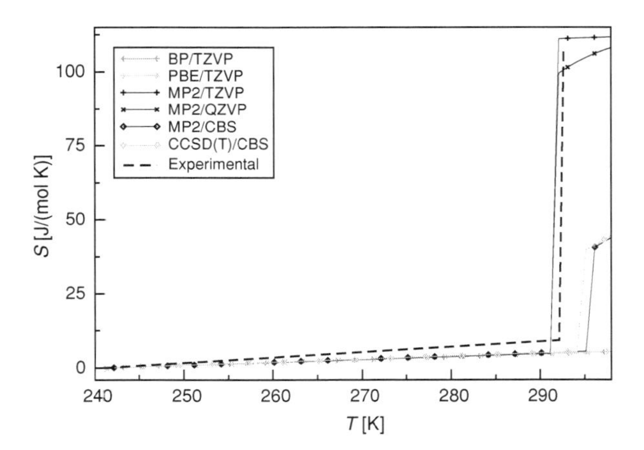

Fig. 5.11 Molar entropy S as a function of the temperature T as obtained from different quantum chemical methods employed for the calculation of the hydrogen fluoride cluster set (see Fig. 4.3) compared to an entropy curve obtained from heat capacity measurements [5]. The reference for all curves is set to $S(241.15\text{ K}) = 0$ J/(mol K). The qce parameters are fitted to reproduce the experimental volume (see Table 4.5). All calculations refer to a pressure of $p = 1 \times 10^5$ Pa

computed for the different approaches, see also Fig. 4.20. In agreement with the isobars computed in the liquid–vapor phase transition domain (see Fig. 5.7), differences start to show up near the phase transition temperature. The entropies obtained from the MP2/TZVP and MP2/QZVP cluster data underestimate the transition temperature by approximately 1 K as in the case of the isobars in Fig. 5.7 and thereby yield the most accurate prediction of the experimental boiling point. However, the corresponding values predicted by the complete basis set interaction energies are of almost equal precision, showing an overestimation of the experimental value of 2 and 3 K, respectively. In complete analogy to the calculated molar volumes, it is seen that the curves based on the dft cluster data do not show a phase transition process at all in the examined temperature range. Considering the phase transition itself it is again apparent that the only accurate prediction of the gas phase entropy can be obtained on the basis of the MP2/TZVP cluster data, for which the phase transition process in the parameter-free model is already complete at the experimental boiling temperature in accordance with Fig. 4.14. As in the case of the molar volume in the gas phase domain (see Fig. 5.8), the gas phase entropies predicted by the remaining MP2 cluster calculations as well as the CCSD(T) cluster data are those of the corresponding parameter-free model, and since the phase transition for these methods is not completely finished in the parameter-free approach (or has not even started in the case of the dft cluster data, see Fig. 4.14), the predicted transition process does not exhibit the qualitative features of the real liquid–vapor phase transition. Thus, the model predictions obtained from the entropy calculations clearly reflect the situation found in the case of the molar volumes and again indicate that a reasonable treatment of the liquid–vapor phase transition in the frame of the qce model is only possible if the corresponding transition process in the parameter-free setup occurs at a lower temperature than the experimental phase transition temperature. It should be noted that the phase transition behavior of the MP2/cbs and the CCSD(T)/cbs data as illustrated in Figs. 5.8 and 5.11 is qualitatively different from the one observed for the MP2/TZVPP pair interaction energies in the case of the water cluster sets (see e.g. Fig. 5.10), which also underestimate the experimental gas phase entropies to a large extent. However, these energies also exhibit an unphysical behavior in the liquid phase temperature interval (as can for instance be seen in the molar volumes plotted in Fig. 5.6) not found in the case of the cbs limit interaction energies for hydrogen fluoride, and more important fail to predict a reasonable liquid–vapor phase transition *even though* the corresponding parameter-free calculation clearly indicates that a realistic gas phase can be obtained from the qce model at the transition temperature, see Fig. 4.6. Consequently, the reason for the inability of the pair interaction energies to yield a physically consistent phase transition process can be attributed to the neglect of cooperativity in these energies, whereas the behavior of the parameter-free model prevents a reasonable description in the case of the MP2/cbs and CCSD(T)/cbs energies of the hydrogen fluoride cluster set.

As has been demonstrated in Sect. 5.1.1 for the case of the parameter-free results, the absolute entropies predicted by the qce model in the liquid–vapor phase transition region can be applied for the calculation of the vaporization entropy in a straightforward way, and the values obtained from the parameter-free model are found to be in the same order of magnitude like the experimental data though at a different transition temperature, see Table 5.1. The same procedure will be employed in the following in connection with the entropies calculated on the basis of the optimized qce parameters in order to predict vaporization entropies at the experimental boiling point. Due to the inability of the qce model to provide a realistic gas phase behavior for the hydrogen fluoride dft and cbs interaction energies at the transition temperature, the linear regression obtained from the high temperature gas phase entropies of the parameter-free approach (see Sect. 5.1.1) are combined with the liquid phase data from the optimized parameter setup for the evaluation of the transition entropy in these cases. For all remaining methods the vaporization entropy is computed directly as the difference between the gas phase value and the liquid phase entropy at the calculated phase transition temperature. The vaporization entropies obtained from this course of action are summarized in Table 5.4 for both water cluster sets and the hydrogen fluoride cluster set.

In addition to the water cluster sets introduced in Sect. 4.1.1, an additional set is considered for the evaluation of the vaporization entropy, namely the optimized spiro set (see the rows "H_2O_{opt}" in Table 5.4). This set is obtained from an optimization procedure of the extended water cluster set, in which cluster structures exhibiting a population of less than 10% are removed from the set if the accuracy of the corresponding isobar in the liquid phase temperature interval is increased by this

Table 5.4 Vaporization entropies at the calculated boiling point as obtained from the optimized parameter setup (see Tables 5.3 (water) and 4.5 (hydrogen fluoride)) for different water cluster sets and the hydrogen fluoride cluster set

Set	dft Methods				
	BP		B3LYP		PBE
	TZVP	TZVPP	TZVP	TZVPP	TZVP
7w8cube	117.7	114.3	119.0	114.0	–
H_2O_{ext}	121.0	118.7	121.3	118.8	–
H_2O_{opt}	121.1	118.8	121.6	119.1	–
HF	113.5	–	–	–	112.7

Set	Post-Hartree–Fock methods					
	MP2					CCSD(T)
	TZVP	TZVPP	pair	QZVP	cbs	cbs
7w8cube	105.9	109.7	72.4	–	–	–
H_2O_{ext}	109.6	112.7	72.4	–	–	–
H_2O_{opt}	109.2	113.0	72.4	–	–	–
HF	106.1	–	–	109.9	108.0	108.0

In the case of the hydrogen fluoride data, the gas phase entropies at the transition temperature are obtained from a linear regression of the parameter-free gas phase values for all methods except MP2/TZVP. All values are given in [J/(mol K)]

elimination [2]. In this way a possible redundancy in the cluster structures and coordination patterns can be corrected via an extended sampling procedure. The cluster set optimized according to this approach contains most of the structures from the extended set, with the exception of the **w3A**, the **s7**, and the **s13** clusters, see Figs. 4.1 and 4.2 [2]. The calculated vaporization entropies in Table 5.4 indicate that the experimental value at the boiling point ($\Delta_{vap}S$(372.76 K) = 109.1 J/(mol K)) is predicted to a reasonable precision by most methods and cluster sets [7]. In almost all cases the experimental value is overestimated in accordance to the same trend observed in the absolute entropies, see Figs. 5.9 and 5.10. This overestimation is most pronounced in the case of the dft cluster data and the larger cluster sets (H_2O ext and H_2O opt). In general, the results obtained from the extended cluster set and the optimized cluster set do not differ by more than 0.5 J/(mol K), thereby indicating that the optimization procedure does not influence the entropic features of the phase transition process to a significant extent. It is also clearly apparent that the increase of the basis set size leads to more accurate values for both employed density functionals and all water cluster sets, and that an analogous convergence towards the experimental result is not obtained if the cluster set is increased instead. The results calculated on the basis of the fully cooperative MP2 cluster data are more accurate as those predicted by the dft methods regardless of the size of the cluster set applied, and the deviations observed for this case do not exceed 4 J/(mol K). Highest accuracy is found for the vaporization entropy based on the MP2/TZVPP cluster data in the case of the 7w8cube cluster set and the values from the MP2/TZVP cluster calculations in the case of the two larger cluster sets. For these combinations the experimental vaporization entropy is predicted almost quantitatively, but it is again seen that this result cannot be obtained by simply combining the largest basis set (TZVPP) and the cluster set yielding the most precise isobar in the liquid phase (H_2O opt). Nevertheless, the results of the liquid phase calculations (see Chap. 4) and of the liquid–vapor phase transition presented so far indicate that a post-Hartree–Fock treatment of the cluster set employing a basis set of at least triple-ζ quality leads to thermodynamic quantities of higher precision as those obtained from corresponding dft cluster data in almost all cases. The vaporization entropies predicted by the MP2/TZVPP pair interaction energies do not show any dependancy on the employed cluster set. In accordance with the corresponding curves in Figs. 5.9 and 5.10, large discrepancies of more than 35 J/(mol K) are obtained with regard to the experimental value, which again indicate that the essential physics of the phase transition process cannot be captured if cooperative effects are neglected. Furthermore, it can be seen that in the case of the phase transition entropy these effects are even more important than the rather strong intermolecular association in liquid water, which leads to the relatively large vaporization entropy of $\Delta_{vap}S$(372.76 K) = 109.1 J/(mol K) as compared to vaporization entropies of typical unassociated liquids. These can be estimated following Trouton's rule to approximately amount to $\Delta_{vap}S \approx 87$ J/(mol K), thereby indicating that the neglect of cooperativity results in even larger discrepancies in the vaporization entropy than the neglect of the association via hydrogen bonding would lead to [1, 13]. A similar effect is found for the phase transition entropies calculated from the parameter-free model (see Table 5.1), which show a higher agreement to the experimental situation than the vaporization entropy calculated from the pair interaction energies in the optimized parameter setup even though no interaction between the different clusters is possible in the parameter-free qce model. Thus, for the determination of the entropy change upon the transition from the liquid to the gas phase the consideration of cooperative effects in the interparticle interactions can be classified as being more important than the accurate treatment of the extended hydrogen bond network present in liquid water.

The vaporization entropies in Table 5.4 calculated for the hydrogen fluoride cluster set show larger deviations to the experimental value at the boiling point ($\Delta_{vap}S(292.69\ \mathrm{K}) = 98.7\ \mathrm{J/(mol\ K)}$) as in the case of the water calculations. The values obtained from the dft cluster data again overestimate the experimental vaporization entropy in a significant way (approximately 15 J/(mol K)), with the largest discrepancy found for the BP/TZVP cluster calculations. As in the case of the phase transition entropies calculated in the frame of the parameter-free model (see Table 5.1), the values predicted by the post-Hartree–Fock methods are lower by several entropic units (J/(mol K)) and thereby are more accurate. The closest agreement to the experimental value is observed for the MP2/TZVP cluster data, which in addition is the only combination capable of predicting a realistic phase transition behavior on the basis of the parameters optimized to the liquid phase molar volume, see Fig. 5.11. Additional inaccuracies of several J/(mol K) could be introduced to the values obtained for the remaining methods in terms of the linear regression analysis of the gas phase entropies calculated from the parameter-free model. However, in the frame of the methodology chosen for the determination of the entropy difference between the gas and the liquid phase the most accurate values are found for the cbs limit cluster interaction energies, and a basis set size effect of approximately 2 J/(mol K) is observed for the transition from the quadruple-ζ basis set (QZVP) to the cbs limit. An additional reason for the larger inaccuracies in the hydrogen fluoride vaporization entropies as compared to those calculated for water is possibly again given by the highly non-ideal character of the real hydrogen fluoride gas phase, for which a considerable population of clusters larger than the monomer has been observed [6]. The presence of these larger aggregates can be expected to have a more pronounced influence on the vaporization enthalpy (as it is clearly seen in the phase transition enthalpies obtained from the parameter-free model, see Sect. 5.1.2), but this behavior possibly leads to a higher degree of ordering in the real vapor phase and thereby to a smaller increase in entropy upon the phase transition. Since these structural details are not captured in the gas phase predicted by the parameter-free qce model, an overestimation of the experimental vaporization entropy can be expected with regard to this point.

The results of the present section demonstrate that reasonable vaporization entropies can be calculated in most cases in the frame of the qce approach either by adjusting the qce parameters to the phase transition point or by simply extending the temperature interval accordingly and employing parameters adjusted to the molar volume of the liquid phase. As in the case of the phase transition isobars discussed in Sect. 5.2.1, the necessary prerequisite for a true first principles prediction of the vaporization entropy lies in the capability of the parameter-free model to yield a reasonable gas phase at the phase transition temperature. If this condition is fulfilled and in addition cooperative effects are considered, a determination of the vaporization entropy to within 10% is possible in the case of water and hydrogen fluoride, with the post-Hartree–Fock cluster data yielding a higher accuracy in general. If no stable gas phase is found at the phase transition point, a linear extrapolation of gas phase entropies from the high temperature domain can be employed for the calculation of the entropy difference, but a loss of accuracy of several entropic units is expected to occur via this approach.

5.2.3 Decomposition into Cluster Degrees of Freedom and Summary

For the assessment of the conventional rrho approach presented in Chap. 3 the examination of the entropy contributions arising from the different molecular degrees of freedom proved to be very instructive and demonstrated that the essential problems of this approach for the calculation of liquid phase entropies are rooted in the volume available for translational motion and the application of the Sackur–Tetrode equation in its unmodified form, see Eq. 3.8. This analysis is possible due to the factorization of the molecular partition function q into the degree-of-freedom dependent contributions according to Eq. 2.26, and from the derivation of the qce approach as presented in Sect. 2.2.1 it is apparent that an equal factorization approach is employed in the case of the different cluster partition functions q_j, see Eq. 2.24. Consequently, an analogous decomposition into the entropy contributions from the various degrees of freedom occurring in each cluster j can be accomplished for the qce entropy as well. In contrast to the decomposition of the entropy in the rrho approach as presented in Chap. 3, the decomposition scheme employed for the translational part in this section does not contain separate contributions from the mass and the number density (as e.g. expressed in Eq. 3.17), but rather differentiates between a real translational contribution S_{trans} depending on the cluster mass as well as the volume available for translational motion according to Eq. 2.49 and a cluster population contribution S_{ind}. This partitioning reflects the contributions relevant for the qce approach in a more detailed way and considers the fact that an analogon to the classical Sackur–Tetrode equation does not exist in the qce model. As pointed out in Sect. 3.2.1, the entropy arising from the $N!^{-1}$ factor occurring in the N-particle partition function (see Eq. 2.20) due to the indistinguishability of the particles is normally included in the translational entropy contribution, thereby leading to the occurrence of the inverse number density in the Sackur–Tetrode equation in combination with the dependancy on the translational volume. However, in the case of the qce model the relevant quantity is not the total particle number N but rather the set of cluster populations $\{n_j\}$ obtained in the course of the iterative procedure. This is most directly seen in the form of the canonical QCE partition function (see Eq. 2.46), which does not depend on the total number of monomers (usually fixed to one mole in a qce calculation), but rather on the various cluster populations n_j. Accordingly, the entropy arising from the cluster population distribution will be treated as a separate contribution to the entropy labeled as S_{ind} (because this contribution is ultimately rooted in the indistinguishability of clusters belonging to the same species j), whereas the translational entropy contribution S_{trans} only depends on the mass and the volume in the present decomposition scheme in agreement with Eq. 2.49.[4] It should be noted that the additive composition of the total entropy from the different contributions at a given temperature is not affected by this new partitioning, i.e., the equality $S_{\text{tot}} = S_{\text{trans}} + S_{\text{rot}} + S_{\text{vib}} + S_{\text{el}} + S_{\text{ind}}$ is still valid [1]. The decomposition of the qce entropy in the liquid–vapor temperature interval of water as obtained on the basis of the MP2/TZVPP cluster calculations and the application of the extended water cluster set is illustrated in Fig. 5.12.[5] In this plot the contribution from each individual degree of freedom is adjusted to a reference temperature of $T = 273.15$ K, i.e., the value calculated for one of the degrees of freedom at a given temperature is reduced by the value of the same degree of freedom obtained at $T = 273.15$ K. According to this procedure negative contributions can occur at higher temperatures if the degree of freedom under consideration exhibits a smaller entropy value than the one calculated for this degree of freedom at $T = 273.15$ K.

The plots in Fig. 5.12 indicate that all degrees of freedom yield similar contributions to the total entropy in the low temperature domain. At moderate temperatures larger differences start to show up, and the contributions from the translational degrees of freedom S_{trans} and the population distribution S_{ind} exhibit largest magnitudes as well as the largest slope, but oppose each other in the sign. This behavior results in an almost linear progression of the total entropy in the liquid phase temperature interval. At the boiling temperature, the contributions from the different degrees of freedom rapidly diverge. It is clearly apparent that no entropy contribution is obtained from the electronic degrees of freedom in either the liquid or the gas phase, which is to be expected due to the exclusive occurrence of closed shell structures in the cluster set under examination. At temperatures larger than the boiling temperature vibrational entropies smaller than those at

[4] The indistinguishability of identical cluster structures is of course no real degree of freedom, but nevertheless affects the canonical partition function (and thereby the entropy) in a non-negligible way, which is the reason why this contribution is treated on an equal footing with the remaining degrees of freedom in the employed decomposition scheme.

[5] Please note that the entropy decomposition predicted by the remaining methods qualitatively agrees to the plots in Fig. 5.12. This is also true in the case of the smaller 7w8cube cluster set.

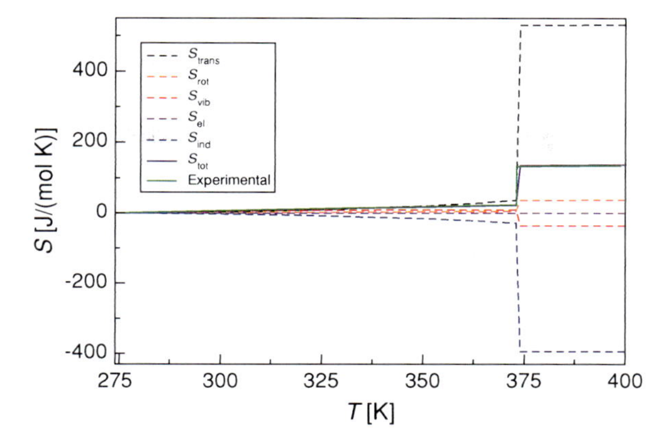

Fig. 5.12 Contributions of the different cluster degrees of freedom to the molar entropy S as a function of the temperature T obtained from the MP2/TZVPP calculations of the extended (7w8cube + spiro) water cluster set compared to the experimental entropies [7]. The reference for all curves is set to $S(273.15\ \mathrm{K}) = 0$ J/(mol K). The qce parameters are fitted to reproduce the experimental volume (see Table 5.3). All calculations refer to a pressure of $p = 1 \times 10^5$ Pa

$T = 273.15$ K are observed according to the negative vibrational values in this temperature domain in Fig. 5.12, whereas a positive vibrational contribution is predicted over the whole liquid phase interval. This result can be rationalized in terms of the computed cluster populations, which indicate that larger cluster structures and higher coordination numbers are present in the low temperature region, whereas the vapor phase is almost exclusively constituted by the monomer structure [1–3]. Due to the higher number of normal modes found in the extended cluster structures especially in the low wavenumber region, the vibrational entropy contribution assumes larger values at low temperatures and thereby leads to a negative contribution in the chosen reference scheme in the gas phase. The same reasoning explains the considerable increase in magnitude observed for the population contribution to the total entropy at the boiling point and in the gas phase. The almost exclusive population of the monomer unit in the qce gas phase results in a significantly smaller entropy contribution as the one obtained from the more homogeneous population distribution in the low temperature domain, and consequently the entropy arising from this factor in the qce partition function drops to approximately −400 J/(mol K) at the boiling point. In contrast, the translational entropy which also assumes the largest values in the liquid phase region steeply increases to approximately 530 J/(mol K) at gas phase temperatures. This considerably enlarged translational entropy in the gas phase can be rationalized in terms of the change in molar volume accompanying the liquid–vapor phase transition (see e.g. Fig. 5.6), which will also affect the free volume of translation to a major extent. In addition, the excluded volume correction (which reduces the translational volume according to Eq. 2.49) exhibits smallest values in this temperature domain, because the monomer unit is almost exclusively populated, and it is apparent that this cluster structure will exhibit the smallest individual cluster volume in the van der Waals sphere model. Consequently, the only significant contribution to the excluded volume term according to Eq. 2.48 in the gas phase will be the excluded volume of the monomer and the calculated V_{ex} value will be considerably smaller as compared to the liquid phase temperature domain. Thus, it is apparent that in the applied decomposition scheme the major factors leading to an entropy increase in the gas phase (larger volume of the cluster phase, smaller excluded volume) are merged into the translational entropy contribution, whereas the most important compensation of this entropy gain in the the vapor phase is found in the uniform population distribution and the corresponding entropy part, with a smaller effect of this particular population set also being visible in the vibrational gas phase entropy. In the case of the rotational entropy contribution a relatively small increase is observed at the liquid–vapor phase transition point, which can again be attributed to the uniform population distribution predicted for the gas phase. In principle, a similar decline of the entropy at the boiling temperature as obtained in the case of the vibrational degrees of freedom could be expected for the rotational entropy as well, since the relevant rotational cluster quantities (the principal moments of inertia) are smaller in the case of the water monomer as compared to the cluster structures exhibiting a relevant population in the low temperature domain. However, this evident effect is again outbalanced by the more homogeneous population distribution calculated in the liquid phase temperature range, which reduces the contribution of each individual cluster structure in a linear way according to Eq. 2.68, whereas possible differences in the principal moments of inertia enter the rotational entropy logarithmically as can e.g. be seen from Eq. 3.9 [1]. Thus, the overall situation obtained for the different entropy contributions of the various cluster degrees of freedom in the frame of the qce model is similar to the one found for the conventional rrho approach (see Sect. 3.2.1) insofar as the most important effects in the gas phase domain are predicted for the volume-dependent quantity (S_{trans}) and the quantity related to the particle number (S_{ind}). The contributions calculated for the rotational and vibrational degrees of freedom affect the total vapor phase entropy to a considerably lesser extent and furthermore exhibit a different sign in the high temperature interval, which results in an almost complete cancellation of the entropy originating from these degrees of freedom in the gas phase. As in the rrho entropy calculations no contribution to the qce entropy is predicted for the electronic degrees of freedom if only closed-shell species are present in the cluster set regardless of the temperature interval considered. The entropy contributions obtained for the liquid phase indicate largest influences of those degrees of freedom related to the volume and the particle number (S_{trans} and S_{ind}) in this temperature domain as well, thereby justifying the free volume approach for the calculation of condensed phase entropy changes in the frame of the rrho model as proposed in Sect. 3.2.3, which essentially neglects possible alterations of the remaining degrees of freedom upon the transition to the liquid phase.

In summary, the result presented in this chapter demonstrate the general possibility of calculating phase transition properties for the liquid–vapor phase transition in the frame of the qce model. The examined systems can be understood as being representative examples of typical associated (hydrogen-bonded) liquids, and therefore a general capability of the qce approach to model realistic liquid–vapor phase transition processes can be expected if some additional specifications as discussed in the present chapter are met. In addition, previous studies demonstrate that the treatment of the solid–liquid phase transition can be realized in the frame of this model as well if cluster structures corresponding to the solid phase are considered in the cluster set [14, 15]. In line with the previous chapters, the most important observations regarding the calculation of liquid–vapor phase transition properties within the qce model will be reiterated in the following.

- According to the results discussed in Chap. 4, the calculation of thermodynamic quantities in the frame of the parameter-free model demonstrates that the consideration of intermolecular interactions in terms of a cluster equilibrium of ideal (non-interacting) cluster structures is sufficient for generating the qualitative features of two distinct phases if the temperature interval is chosen large enough. A linear regression analysis of the thermodynamic quantities calculated for the low density gas phase in the parameter-free approach can be used for the extrapolation of these quantities to lower temperatures at which the liquid-like phase is predicted to be stable, see Figs. 5.1 and 5.2. The phase transition entropies and enthalpies calculated in this way for the extended water cluster set are in the same order of magnitude as the corresponding vaporization properties even if the real phase transition process occurs at considerably larger temperatures, see Tables 5.1 and 5.2. Larger deviations are observed in the case of the hydrogen fluoride cluster set especially in the case of the phase transition enthalpy, which can be attributed to the highly non-ideal character of the hydrogen fluoride vapor phase. These anomalies are not captured in the qce results.
- The calculation of specific heat capacities in the parameter-free setup shows clear indications of a phase transition process, see Fig. 5.3. The transition temperatures predicted by the turning points in the heat capacities significantly overestimate the experimental phase transition temperature in the case of the dft cluster data of the hydrogen fluoride cluster set, whereas only a small overestimation is observed for the MP2/cbs and CCSD(T)/cbs hydrogen fluoride cluster data and an underestimation is found for the MP2/QZVP and MP2/TZVP cluster calculations. On the contrary, the phase transition temperatures obtained in this way for both water cluster sets underestimate the experimental boiling point by more than 100 K regardless of the quantum chemical method employed for the cluster calculations. The comparison to experimental data shows a reasonable agreement of the calculated gas phase heat capacity in the case of hydrogen fluoride, whereas the heat capacities predicted for the liquid-like phase by the parameter-free model are smaller than those obtained for the gas phase and thereby exhibit a qualitatively wrong behavior.

- The readjustment of the qce parameters with regard to the extended temperature interval leads to a reasonable and qualitatively correct reproduction of the volumetric properties of the phase transition process for both water cluster sets and most methods employed for the cluster calculations, see Figs. 5.4 and 5.6. In contrast to the parameter adjustment results of the liquid phase, the new parameters are identical for both cluster sets and all employed methods, see Table 5.3. This observation can be traced back to the applied parameter adjustment procedure, which evaluates the volume difference between the qce isobar and the experimental reference at each temperature point and therefore favors those isobars predicting the boiling point correctly. In a given parameter interval this exact reproduction is only possible for a certain parameter configuration not depending on the employed cluster data.
- In contrast to the fully cooperative cluster calculations the behavior of the isobar computed from the MP2/TZVPP pair interaction energies is found to be qualitatively wrong in both water cluster sets, see Figs. 5.4 and 5.6. The possibility to correct the missing fraction of intermolecular interaction via an increased intercluster mean field contribution as found in the case of the liquid phase calculations (see e.g. Table 4.4) is not given in the extended temperature setup if the phase transition point is to be reproduced correctly.
- The simple extension of the temperature interval in the case of the hydrogen fluoride cluster set indicates that the parameters solely adjusted to the liquid phase can account for a phase transition process to within 4 K of the experimental boiling point, see Fig. 5.7. However, a qualitatively correct behavior of the phase transition can only be predicted if the parameter-free model exhibits a stable vapor phase at the experimental boiling point to which the transition can occur. In the case of the hydrogen fluoride qce calculations, this is only given for the MP2/TZVP cluster data, see Fig. 4.7. This observation can be expected to be a general feature of liquid–vapor phase transition processes in the frame of the qce model.
- The application of the readjusted parameters for the entropy in the liquid–vapor phase transition domain yields a reasonable prediction of both the liquid phase entropy and the gas phase entropy for both water cluster sets and all fully cooperative cluster calculations, see Figs. 5.9 and 5.10. Identical trends as in the case of the liquid phase entropy calculations are observed, see Figs. 4.18 and 4.19. The results obtained from the MP2/TZVPP pair interaction energies exhibit a qualitatively correct progression, but considerably underestimate the experimental entropies in both phases. For the hydrogen fluoride cluster set the only reasonable prediction of the entropy in the liquid as well as in the gas phase is found for the MP2/TZVP cluster data as in the case of the molar volume, see Fig. 5.11, which can again be attributed to the results obtained from the corresponding parameter-free calculations.
- The calculation of vaporization entropies by either taking the direct difference between the absolute values from the vapor phase and the liquid phase or by extrapolating the high temperature gas phase of the parameter-free model down to the experimental boiling point and subsequently calculating the

corresponding difference results in a reasonable prediction of the experimental values with deviations not larger than 10%, see Table 5.4. The most accurate results are found for the fully cooperative MP2 calculations of the different water cluster sets, whereas the vaporization entropies calculated from the pairwise interaction energies only recover 2/3 of the experimental value. This deviation is larger than the one found from the estimation of the vaporization entropy according to Trouton's rule ($\Delta_{vap}S \approx 87$ J/(mol K)), which indicates that the consideration of cooperativity is more important than the extended association network occurring in liquid water in the case of the vaporization entropy.

- The decomposition of the calculated entropy in the liquid–vapor domain exemplarily carried out for the results obtained from the MP2/TZVPP calculations of the extended water cluster set yields largest but opposing contributions from the translational degrees of freedom (depending on the volume) as well as from the population distribution, see Fig. 5.12. A uniform population of a single cluster structure results in an entropy decrease as compared to a more homogeneous distribution. This observation is an important factor for the calculated vapor phase entropies, in which the water monomer is almost exclusively populated and which is necessary to compensate for the large translational contribution arising from the increased volume at these conditions. Less significant effects are found for the rotational and vibrational entropy contributions both in the vapor phase as well as in the liquid in accordance with the conclusions of Chap. 3, thereby again demonstrating that the proper treatment of the accessible volume is the most urgent matter for the calculation of condensed phase entropies in the frame of the rrho model and the qce approach.

From a technical point of view the most important result of the present chapter lies in the observation that the capability of the qce approach to predict the liquid–vapor phase transition in a qualitatively correct way can directly be assessed from a parameter-free calculation for the system under investigation. According to this finding the possibility of calculating the corresponding phase transition properties for a novel compound or cluster set can thus be determined within minutes and does not have to rely on a time-consuming parameter adjustment procedure. If the conditions for a stable gas phase are predicted by the parameter-free setup, the calculation of vaporization entropies seems to be possible to a reasonable precision in the frame of the qce approach, and the results listed in Table 5.4 indicate that the obtained results are not significantly affected by variations in the cluster set. Thus, accurate results can already be obtained from small cluster sets like the 7w8cube set if post-Hartree–Fock methods in combination with basis sets of adequate size (at least triple-ζ quality) are employed for the cluster calculations. Considering the fact that quantities like the vaporization entropy can be obtained from other atomistic approaches only at relatively large difficulties, the simple and straightforward procedure as introduced in the present chapter could provide a helpful addition to established methods, e.g., relying on md simulation techniques [16–18].

References

1. Spickermann C, Lehmann SBC, Kirchner B (2008) J Chem Phys 128:244506
2. Lehmann SBC, Spickermann C, Kirchner B (2009) J Chem Theory Comput 5:1640–1649
3. Lehmann SBC, Spickermann C, Kirchner B (2009) J Chem Theory Comput 5:1650–1656
4. Vanderzee CE, Rodenburg WW (1970) J Chem Thermodyn 2:461–478
5. Yaws CL (eds) (1999) Chemical properties handbook. McGraw-Hill, New York
6. Janzen J, Bartell LS (1969) J Chem Phys 50:3611–3618
7. National Institute of Standards and Technology, "NIST chemistry webbook", see http://webbook.nist.gov/
8. Weinhold F (1998) J Chem Phys 109:373–384
9. Perlt E (2008) Isochore Wärmekapazität und Quantum Cluster Equilibrium-Methode. Thesis, Universität Leipzig
10. Kirchner B (2007) Phys Rep 440:1–111
11. Kirchner B (2005) J Chem Phys 123:204116
12. Simons JH, Bouknight JW (1932) J Am Chem Soc 54:129–135
13. Trouton F (1884) Phil Mag 18:54–57
14. Ludwig R, Weinhold F (1999) J Chem Phys 110:508–515
15. Lenz A, Ojamäe L (2009) J Chem Phys 131:134302
16. Frenkel D, Smit B (2002) Understanding molecular simulations. Academic Press, San Diego
17. Peter C, Oostenbrink C, van Dorp A, van Gunsteren WF (2004) J Chem Phys 120:2652–2661
18. Panagiotopoulos AZ (1987) Mol Phys 61:813–826

Chapter 6
Outlook

As already pointed out in the introduction, the major objective of this thesis is the investigation of methods and approaches for the calculation of the entropy on the basis of well-established quantum mechanical models such as Kohn–Sham density functional theory and Møller–Plesset perturbation theory, i.e., on the basis of the first principles of quantum mechanics. In contrast to "mechanical" properties like e.g. the dipole moment, the entropy is a thermodynamic property and therefore not in direct reach of the methods routinely applied in quantum chemistry. This fact does not constitute a major problem as long as the system under investigation essentially exhibits a behavior which is close to the ideal gas state, i.e., as long as interparticle interactions and excluded volume effects can be neglected as a first approximation. This assumption is valid for most compounds in the gas phase. In these cases the statistical thermodynamics of the ideal gas can be incorporated into the quantum chemical calculations in a most straightforward way, and most of the modern quantum chemical program packages routinely provide thermodynamic quantities as a part of their output [1–3]. This merge of quantum chemistry and ideal gas statistical thermodynamics has become known as the rigid rotor harmonic oscillator model, and due to the unproblematic availability of the results via quantum chemical program packages nowadays is employed in a variety of applications in all fields of chemistry, see for instance the literature review parts of Sects. 3.2.1 and 3.2.3. However, in some cases the foundations of this approach seem to have been forgotten. Problems start to appear for systems no longer adequately treatable in the frame of the ideal gas approximation, which in most cases arises due to an increased complexity in the intermolecular interactions or simply due to the density of the system at the chosen external conditions. This is the point where the investigations of the present study become relevant.

The studies undertaken in Chap. 3 concentrate on the behavior of the conventional approach at those conditions it was originally not developed for, i.e., the environmental effects are simply ignored in the investigations of Sect. 3.1 (with the exception of the microsolvation approach expressed in Eq. 3.2).

C. Spickermann, *Entropies of Condensed Phases and Complex Systems*, Springer Theses, DOI: 10.1007/978-3-642-15736-3_6,

The comparison of this most straightforward procedure to experimental results of the association reaction between supramolecular compounds in solution demonstrates that the problematic contribution to the free energy change lies in the calculated entropy change, which shows an extraordinary dependancy on the stoichiometry of the underlying reaction. A detailed examination of the entropy calculation in the rrho approach as well as a quantitative error analysis ultimately indicate that this dependancy stems from the accessible volume in the translational entropy contribution. Similar inconsistencies in the translational entropy change of condensed phase association/dissociation reactions have been observed before in the biochemical sciences and are still a subject of great controversy, see, e.g., Ref. [4]. In the frame of the rrho model a simple volume scaling approach originally suggested by Amzel et al. has been found to considerably improve the situation and yield deviations between the calculated and measured free energy changes of less that 5 kJ/mol in most cases (see Sect. 3.2.3) [5]. Thus, it could be shown that a simple modification is sufficient for an approximate adaption of the original rrho approach to the high density domain in the case of the investigated pseudorotaxane systems, but a larger test set systematically covering different kinds of intermolecular interactions as well as a larger number of structural (and chemical) motifs certainly has to be examined in order to give a definite assessment of the proposed approach. However, from the viewpoint of a first principles prediction of the entropy the free volume model employed in Chap. 3 introduces the additional difficulty of calculating (or estimating) the free volume v_f of a particle in solution. A numerical evaluation of the localized configurational integral (see Eq. 3.34) is certainly possible on the basis of sampling techniques like md or Monte Carlo simulations (as has been demonstrated in Ref. [5] for several liquids), but larger complications arise in the case of static first principles calculations, for which the proposed modification of the rrho model actually is intended. In order to overcome this problem alternative routes towards the determination of (free) volumes on the molecular scale could prove to be useful, for instance by combining appropriate cluster calculations with a geometrical determination of the free volume in the optimized cluster structure e.g. in terms of a Voronoi polyhedra analysis [6, 7]. If this is not possible for any reason, a practical alternative is probably given by microsolvation approaches as those suggested in Eq. 3.2. According to the results of the corresponding entropy calculations in Chap. 3, the major intent of the additional solvent molecules should lie in the artificial generation of a model reaction not exhibiting any particle effect, i.e., in which no interconversion between different types of degrees of freedom occurs. Such a course of action will not result in a large increase of the computational effort in most cases and will not be affected by the artificially large volume contributions to the translational entropy due to the constant number of translational degrees of freedom on both sides of the reaction arrow, which leads to a cancellation of these effects in the calculation of the reaction entropy. If the system size is not too large and an even higher accuracy in the calculated thermodynamic reaction quantities is required, approaches more sophisticated than the original rrho model like e.g. the full vibrational configurational interaction method for the calculation of vibrational contributions or the consideration of hindered rotations via numerical methods can be suggested [8, 9].

An approach closely related to the concepts of the rrho model is constituted by the quantum cluster equilibrium theory, which essentially considers a fraction of the intermolecular interactions on a local scale by employing distinct cluster structures as the fundamental units instead of the isolated molecule routinely applied in static quantum chemical calculations. A detailed analysis of this model concerning the reproduction and prediction of thermodynamic liquid phase properties as well as the corresponding properties of the liquid-vapor phase transition is presented in Chaps. 4 and 5 for liquid water and liquid hydrogen fluoride, which can be expected to be generic examples of hydrogen-bonded fluids. An important result of these studies lies in the observation that the consideration of a single isolated cluster structure (i.e., a single cluster partition function q_j) is not sufficient for generating a qualitatively correct liquid phase behavior in thermodynamic terms. Instead, the findings presented in Sec. 4.1.2 demonstrate that the essential feature for the transition from the ideal gas state to a liquid-like behavior is the establishment of an equilibrium between the different cluster structures and the consideration of the corresponding canonical partition function Q, even if this equilibrium is formulated in terms of the ideal gas state (i.e., the clusters are not allowed to interact with each other and do not exhibit any kind of spatial volume). The condensation process observed in the parameter-free calculations clearly indicate that the consideration of a thermodynamic equilibrium between clusters of varying size captures the qualitative features of a low density gas phase and a liquid-like phase as well as the transition between these two states. If no exceptional circumstances (like the vapor phase association of hydrogen fluoride) are present, the enthalpy and entropy changes calculated for the transition between these two model phases are in the same order of magnitude as the experimental values, i.e., an order-of-magnitude estimate of phase transition properties from first principles should in general be possible on the basis of this approach. A more accurate determination of liquid phase properties as well as phase transition properties can be realized by modifying the ideal gas cluster equilibrium in terms of a van der Waals-like extension. As in the case of the classical model of the van der Waals gas, this extension introduces a parameter for the mean field interaction between the cluster structures as well as a parameter for the volume excluded due to the combinded volume of the cluster structures to the model, which in the present formulation of the model cannot be obtained from first principles calculations in a straightforward way. The parameter adjustment procedure employed in this study constitutes a possibility to evaluate these parameters if experimental data is available in the temperature range of interest, but methodological improvements in this part of the model would clearly be helpful. The qce parameters optimized with regard to the experimental isobars of water and hydrogen fluoride show non-negligible variations between these two compounds, but differences arising from variations in the cluster set, in the temperature interval, or in the quantum chemical methodology applied for the cluster calculations are less pronounced. Thus, the existence of substance-specific sets of "universal" parameters seems likely to which the qce results might converge if the cluster set and the quantum chemical methodologies are improved, and the determination of these parameters on the basis of first principles methods will certainly be an important issue for the improvement of the predictive capabilities of the qce approach. In the frame of the present model it is seen e.g. from the results in Sects. 4.2.2 and 5.2.2 that the parameters obtained from the adjustment to experimental isobars in most cases predict liquid phase entropies and vaporization entropies of reasonable accuracy if two additional prerequisites are met. The first of these conditions is the existence of a stable vapor phase in the parameter-free model at the experimental boiling temperature to which the phase transition in the optimized parameter calculation can occur, and the second one is the consideration of cooperative effects in the supramolecular cluster interaction energies. The impact of these cooperative effects on the modeling of the phase transition has been observed to be of such extent that a reasonable behavior can neither be computed for the liquid phase temperature interval nor for the phase transition itself if only pairwise interaction energies are considered and the boiling temperature is to be predicted exactly. To further illustrate this point the vaporization entropy predicted by the pairwise additive approach has been compared to the estimate according to Trouton's rule for completely non-associated liquids, and is is found that the discrepancies arising from the neglect of cooperativity are even more severe than the neglect of association in liquid water simulated in this way. This observation is possibly of high significance concerning the treatment of the solid-liquid phase transition in the frame of the qce model. Depending on the substance under investigation, considerably larger cluster structures could be required for a consistent modeling of this phase transition process, and it is apparent from the results presented here that such structures on all accounts have to be treated in a fully cooperative fashion and not e.g. in terms of an empirical pair potential if a reasonable result is to be expected.

Even though the results of the present study are encouraging, is is also apparent that a lot of effort still has to be put into the development of the qce approach. The possibly most important point is the elimination of the arbitrary element in the construction of the cluster set, which in the present model is either designed according to chemical intuition or employing experimental information e.g. about coordination patterns relevant in the system under investigation. A possible alternative to the present situation could lie in a preconditioning of the cluster set e.g. in terms of a Monte Carlo sampling of differently sized initial cluster structures. For this process the application of Kohn–Sham density functional theory in combination with basis sets of small to moderate size (or even an empirical force field appropriate to the examined system) might be sufficient. The most stable structures identified by this sampling procedure could afterwards be subjected to a more sophisticated treatment in terms of high quality ab initio approaches. A prevalent observation apparent from most results in this thesis is an increased accuracy of those qce results which are based on high quality quantum chemical calculations. In general, thermodynamic quantities obtained from the MP2 (or the CCSD(T)) cluster data exhibit a higher precision as compared to the corresponding dft results if the applied basis set is of sufficient size, which for the systems investigated can be estimated to amount to triple-ζ quality with consideration of additional polarization functions (i.e., TZVPP). A notably exception to this behavior is constituted in the failure of both the MP2/cbs and the CCSD(T)/cbs cluster calculations (which represent the most accurate quantum chemical approaches employed in this thesis) to predict a qualitatively correct phase transition process for hydrogen fluoride. Nevertheless, the progression of the liquid phase molar volume in combination with the predicted phase transition *temperature* is found to be of highest precision in the case of the CCSD(T)/cbs cluster data, and the inability to model a reasonable phase transition is clearly rooted in the parameter-free qce calculation. Thus, the calculation of liquid phase thermodynamic quantities for substances exhibiting a demanding electronic structure on the basis of high quality ab initio cluster computations can be expected to be successful, which in contrast could give rise to larger problems in the frame of alternative approaches relying e.g. on dft methods or empirical force fields. Chemical substances of this kind can therefore be expected to constitute an important field for future applications of the qce approach.

References

1. Ahlrichs R, Bär M, Häser M, Horn H, Kölmel C (1989) Chem Phys Lett 162:165–169
2. Neugebauer J, Reiher M, Kind C, Hess BA (2002) J Comput Chem 23:895-910
3. Frisch MJ et al (2004) Gaussian03
4. Zhou HX, Gilson MK (2009) Chem Rev 109:4092–4107
5. Siebert X, Amzel LM Proteins (2004) 54:104–115
6. Okabe A (2000) Spatial tesselations: concepts and applications of Voronoi diagrams. Wiley, New York
7. Schröder C, Neumayr G, Steinhauser O (2009) J Chem Phys 130:194503
8. Roy TK, Prasad MD (2009) J Chem Phys 131:114102
9. Wilson EB Jr (1959) Adv Chem Phys 2:367–393

Chapter 7
Appendix

7.1 Computational Details

All quantum chemical calculations have been performed employing the TURBOMOLE 5.91 program package with the exception of the coupled cluster results (obtained from the MOLPRO package) as well as the natural bond orbital analysis (obtained from the GAUSSIAN suite of programs) [1–3]. The second order Møller–Plesset perturbation theory (MP2) and the gradient-corrected density functional theory results have been obtained in the frame of the resolution of identity (ri) approximation [4, 5]. In all calculations the energy convergency criterion was fixed to 10^{-8} a.u. and the maximum norm of the cartesian gradient has been constrained to 10^{-4} a.u. in all cases, with the exception of the hydrogen fluoride MP2/QZVP$^{☆}$ results presented in Chap. 4, for which it is decreased to 10^{-5} a.u. The complete basis set studies employed in the frame of the MP2 and the CCSD(T) approaches have been obtained from single point calculations on the structures predicted by the tight convergency MP2/QZVP$^{☆}$ geometry optimizations. In these cases the incremental scheme had to be applied for the larger hydrogen fluoride cluster structures in order to keep the single point calculations feasible from a computational point of view [6, 7]. The basis sets employed for the complete basis set limit extrapolations are the Dunning basis sets aug-cc-pVTZ and aug-cc-pVQZ, whereas all remaining quantum chemical calculations have been obtained from the Ahlrichs basis sets TZVP, TZVPP, or QZVP [1, 8–11]. The extrapolation to the basis set limit has been performed in terms of the two point formula of Halkier and Helgaker et al. [12].

For the determination of the harmonic frequencies the SNF program has been used in all cases, which computes the wave numbers of the normal modes in the frame of the harmonic approximation as numerical derivatives of the analytic gradients calculated by the structure optimization routine [13]. All optimized geometries have been checked to be true minima on the respective potential energy surfaces by examining the corresponding harmonic wave numbers. The only

C. Spickermann, *Entropies of Condensed Phases and Complex Systems*,
Springer Theses, DOI: 10.1007/978-3-642-15736-3_7,

structures exhibiting one (or more) negative eigenvalues are hydrogen fluoride chain clusters containing more than three monomer units in the case of the MP2/QZVP combination. These structures are therefore not considered in the qce calculations. In almost all cases evidence was found that the three-point formula for the numerical differentiation is of sufficient accuracy, but in some cases the seven-point formula had to be applied in order to obtain a complete set of positive wave numbers [14]. All thermodynamic quantities based on the conventional rrho model have been obtained from the SNF standard output files. For the dft calculations, the following density functionals were used throughout this thesis:

- BP [15, 16]
- PBE [17]
- B3LYP [18, 19]

The calculated supramolecular interaction energies (see Eq. 2.50) have been approximately corrected for the basis set superposition error (bsse) in terms of the generalization of the counterpoise correction as introduced by Wells and Wilson [20, 21]. No bsse correction scheme has been applied to the interaction energies calculated at the cbs limit.

7.1.1 Rigid Rotor Harmonic Oscillator Calculations (Chap. 3)

All calculations have been carried out employing the gradient-corrected density functional BP and the TZVP basis set [10, 11, 15, 16]. The two-center shared electron number (SEN) employed in the calculation of the individual hydrogen bond energies (see Tables 3.3 and 3.5) as well as for the determination of atomic charges (see Table 3.4) are based on the Davidson population analysis [22, 23]. The SEN analysis is based on a linear relation between the hydrogen bond energy E^{SEN} and the corresponding two-center shared electron number σ between the donor hydrogen atom and the acceptor atom according to

$$E^{\mathrm{SEN}} = m \times \sigma + b, \tag{7.1}$$

from which the individual hydrogen bond energy can be estimated if the slope m as well as the axis intercept b of the linear equation are known. In general, these quantities depend on the acceptor atom of the hydrogen bond, and in accordance with previous studies of amide-type hydrogen bonds in the frame of the BP/TZVP combination values of $m = -724.00$ kJ/(mol e) and $b = 2.01$ kJ/mol have been chosen for these parameters [24].

The calculation of the thermodynamic gas phase reaction data for the association of the pseudorotaxane complexes as well as the dimerization of nitrogen dioxide, acetic acid, and water, have been carried out on the basis of the rrho model as introduced in Sect. 2.1.2. In the case of the gas phase dimerization reactions (see Eqs. 3.4–3.6), the required molecular data has been obtained from quantum chemical calculations employing the BP functional in combination with the TZVPP basis set.

7.1.2 Quantum Cluster Equilibrium Calculations (Chaps. 4, 5)

All quantum cluster equilibrium calculations in this thesis have been performed employing the PEACEMAKER code [25]. The damping parameters for the qce calculations have been set to 0.1 for the damping in the pressure-temperature loop and to 1.0 for the damping in the iteration and the damping of the solution, respectively. The criterion for the convergency of the qce iterations has been fixed to 10^{-6} L in all calculations and the temperature step has been set to 1 K in general. In some cases (MP2/TZVP in combination with the parameter-free model) it was found to be necessary to increase the temperature step to 2 K for a significantly improved convergence behavior. If the qce iterations were found not to converge at single temperature points in the investigated interval, these temperatures were considered in the plots of the calculated molar volumes (see e.g. B3LYP/TZVP in Fig. 4.6), but were eliminated for the calculation of other thermodynamic data, since these can only be obtained from a converged qce partition function in a reasonable way. In the case of the hydrogen fluoride cluster set, increasing convergency problems have been encountered in the lower part of the liquid phase temperature interval. For this reason only the high temperature domain of liquid hydrogen fluoride (241.15–291.15 K) has been examined in the frame of the qce calculations. The initial parameter ranges for the adjustment procedure have been set to 0.1 (J $\times$ m^3/mol)–1.0 (J $\times$ m^3/mol) (a_{mf}) and 0.5–1.5 (b_{xv}). If the optimized parameters were found to be equal to these limiting values, the interval has been adjusted accordingly for the subsequent parameter sampling. The final precision for the evaluation of the qce parameters has been fixed to 10^{-4} (J $\times$ m^3/mol) in the case of the mean field interaction parameter (a_{mf}) and to 10^{-3} for the excluded volume factor (b_{xv}).

7.2 The Effect of the Boltzmann Operator on a Product Wave Function for Non-Interacting Particles

Insertion of Eq. 2.11 into the N-particle partition function Eq. 2.10 yields

$$\begin{aligned} Q_{\mathrm{qm,ideal}} &= \sum_j \left\langle \psi_{j,1}\cdots\psi_{j,N} \middle| \exp\left(-\beta\sum_{i=1}^{N}\hat{h}_i\right) \middle| \psi_{j,1}\cdots\psi_{j,N} \right\rangle \\ &= \sum_j \left\langle \psi_{j,1}\cdots\psi_{j,N} \middle| \left(\sum_{n=1}^{\infty}\frac{(-\beta)^n}{n!}\left[\sum_{i=1}^{N}\hat{h}_i\right]^n\right) \middle| \psi_{j,1}\cdots\psi_{j,N} \right\rangle, \end{aligned} \tag{7.2}$$

where the MacLaurin expansion of the exponential operator is employed again. The result of the operator $(\sum_i^N \hat{h}_i)^n$ on a product wave function $|\psi_{j,1}\ldots\psi_{j,i}\ldots\psi_{j,N}\rangle$ can be made plausible by exemplarily considering the case for $n = 2$:

$$\begin{aligned}
\left[\sum_{i=1}^{N}\hat{h}_i\right]^2 |\psi_{j,1}\ldots\psi_{j,i}\ldots\psi_{j,N}\rangle &= \left(\hat{h}_1+\hat{h}_2+\cdots+\hat{h}_N\right) \\
&\quad\times\left(\hat{h}_1+\hat{h}_2+\cdots+\hat{h}_N\right)|\psi_{j,1}\ldots\psi_{j,i}\ldots\psi_{j,N}\rangle \\
&= \left(\hat{h}_1+\hat{h}_2+\cdots+\hat{h}_N\right)\left(\hat{h}_1|\psi_{j,1}\ldots\psi_{j,i}\ldots\psi_{j,N}\rangle\right. \\
&\quad\left.+\hat{h}_2|\psi_{j,1}\ldots\psi_{j,i}\ldots\psi_{j,N}\rangle+\cdots+\hat{h}_N|\psi_{j,1}\ldots\psi_{j,i}\ldots\psi_{j,N}\rangle\right) \\
&= \left(\hat{h}_1+\hat{h}_2+\cdots+\hat{h}_N\right)\left(\epsilon_1|\psi_{j,1}\ldots\psi_{j,i}\ldots\psi_{j,N}\rangle\times\right. \\
&\quad\left.+\,\epsilon_2|\psi_{j,1}\ldots\psi_{j,i}\ldots\psi_{j,N}\rangle+\cdots+\epsilon_N|\psi_{j,1}\ldots\psi_{j,i}\ldots\psi_{j,N}\rangle\right) \\
&= \left(\hat{h}_1+\hat{h}_2+\cdots+\hat{h}_N\right)\left[\sum_{i}^{N}\epsilon_i\right]|\psi_{j,1}\ldots\psi_{j,i}\ldots\psi_{j,N}\rangle \\
&= \left[\sum_{i}^{N}\epsilon_i\right]^2|\psi_{j,1}\ldots\psi_{j,i}\ldots\psi_{j,N}\rangle
\end{aligned} \tag{7.3}$$

References

1. Ahlrichs R, Bär M, Häser M, Horn H, Kölmel C (1989) Chem Phys Lett 162:165–169
2. Frisch MJ et al (2004) Gaussian03
3. Werner H-J, Knowles PJ, Lindh R, Schütz M et al (2006) 'MOLPRO, version 2006, a package of ab initio programs, see http://www.molpro.net
4. Haase F, Ahlrichs R (1993) J Comput Chem 14:907–912
5. Eichkorn K, Treutler O, Öhm H, Häser M, Ahlrichs R (1995) Chem Phys Lett 240:283–290
6. Stoll H (1992) Chem Phys Lett 191:548–552
7. Friedrich J, Perlt E, Roatsch M, Spickermann C, Kirchner B (2010) J Chem Theory Comput (submitted)
8. Dunning TH (1989) J Chem Phys 90:1007–1023
9. Kendall RA, Dunning TH, Harrison RJ (1992) J Chem Phys 96:6796–6806
10. Schäfer A, Horn H, Ahlrichs R (1992) J Chem Phys 97:2571–2577
11. Schäfer A, Huber C, Ahlrichs R (1994) J Chem Phys 100:5829–5835
12. Halkier A, Helgaker T, Jørgenson P, Klopper W, Koch H, Olsen J, Wilson AK (1998) Chem Phys Lett 286:243–252
13. Neugebauer J, Reiher M, Kind C, Hess BA (2002) J Comput Chem 23:895–910
14. Neugebauer J, Herrmann C, Reiher M (2007) SNF 4.0. ETH Zürich, Zürich
15. Becke AD (1988) Phys Rev A 38:3098–3100
16. Perdew JP (1986) Phys Rev B 33:8822–8824
17. Perdew JP, Burke K, Ernzerhof M (1996) Phys Rev Lett 77:3865–3868
18. Becke AD (1993) J Chem Phys 98:5648–5652
19. Stephens PJ, Devlin FJ, Chabalowski CF, Frisch MJ (1994) J Phys Chem 98:11623–11627
20. Boys SF, Bernardi F (1970) Mol Phys 19:553–566

21. Wells BH, Wilson S (1983) Chem Phys Lett 101:429–434
22. Davidson ER (1967) J Chem Phys 46:3320–3324
23. Roby KR (1974) Mol Phys 27:81–104
24. Reckien W, Eggers M, Vögtle F, Schalley CA, Peyerimhoff SD, Kirchner B to be published
25. PEACEMAKER V 1.4 Copyright B. Kirchner, written by Kirchner B, Spickermann C, Lehmann SBC, Perlt E, Uhlig F, Langner J, Domaros Mv, Reuther P (2004–2009), University of Bonn, Institute of Physical and Theoretical Chemistry, University of Leipzig, Wilhelm–Ostwald Institute of Physical and Theoretical Chemistry Bonn-Leipzig 2009, see also http://www.uni-leipzig.de/ ~ quant/index.html/

Index